Plenty of Room for Biology

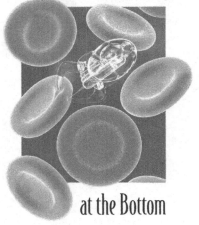

at the Bottom

An Introduction to Bionanotechnology

Plenty of Room for Biology

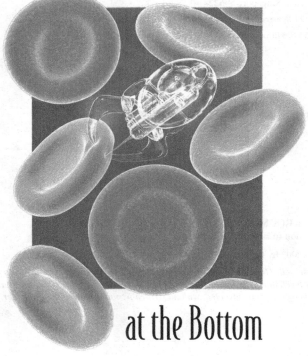

at the Bottom

An Introduction to Bionanotechnology

Ehud Gazit

Tel Aviv University, Israel

Imperial College Press

ICP

Published by

Imperial College Press
57 Shelton Street
Covent Garden
London WC2H 9HE

Distributed by

World Scientific Publishing Co. Pte. Ltd.
5 Toh Tuck Link, Singapore 596224
USA office: 27 Warren Street, Suite 401-402, Hackensack, NJ 07601
UK office: 57 Shelton Street, Covent Garden, London WC2H 9HE

British Library Cataloguing-in-Publication Data
A catalogue record for this book is available from the British Library.

PLENTY OF ROOM FOR BIOLOGY AT THE BOTTOM
An Introduction to Bionanotechnology

ISBN-13 978-1-86094-677-6
ISBN-10 1-86094-677-1

Printed in Singapore.

This book is dedicated to my family with love

Contents

Preface

This book represents a great challenge: It introduces the readers to the integration of two of the most important disciplines of the 21st century, biotechnology and nanotechnology. A special effort was made to expose readers from diverse backgrounds to the basic principles, techniques, applications, and great prospects at the intersection of these two key fields. All of this is done without assuming a strong background in either of the two subjects. We will discuss two related fields: "nanobiotechnology" which relates to the application nanotechnological principles and tools to biology, but also "bionanotechnology" which concerns the use of biological principles such as recognition and assembly for nanotechnological applications.

While the 20th century was the century of physics, electronics, and communication, the years to come are considered to be mainly dominated by the biological revolution that already started in the second half of the 20th century as well as nanotechnology, a new and exciting new front that deals with minuscule nanometer-scaled assemblies and devices. Traditionally, miniaturization was a process in which devices became smaller and smaller by continuous improvement of existing techniques. Yet, we rapidly approach the limits of miniaturization using the conventional top-down fabrication tools. One of the key futuristic ways of making tiny machines will be to scale them up from the molecular level. For that purpose, many lessons could be learned about the arrangement of nano-scale machines as occurs in each and every biological system. The living cell is actually the only place in which genuine functional molecular machines, as often described in futuristic presentations of nanotechnology, could actually be found. Molecular motors, ultrasensitive nano-scale sensors, DNA replication machines, protein synthesis machines, and many other miniature devices exist even in the very simple early pre-bacterial cells that evolved more than 3 billion years ago. When we climb higher on the evolution tree, the nano-machines of course become more sophisticated and powerful. Yet, only billions of years after life emerged, we are beginning to utilize the concepts of recognition and assembly that lead to the formation of nano-scale machines for our technological needs. On the other hand many principles and applications of nano-technology that were developed for non-biological systems could be very useful for immediate biological applications such as advanced sensors and molecular scaffolds for tissue engineering as well as long-term prospects such as *in situ* modifications at the protein and DNA levels.

While the fields of nanobiotechnology and bionanotechnology are very new their prospects are immense. The marriage between biotechnology and nanotechnology could lead to a dramatic advancement in the medical sciences. It may well be a place in which many of the current diseases and human disorders will be eradicated. In a reasonable time-scale, cancer and AIDS may be regarded in the same way that polio and tuberculosis are being considered now. Genetic defects could be identified and corrected already even before birth. Nano-scale robots that may be inserted into our body could perform very complicated surgical tasks such as a brain surgery. Nano-machines may even be able to solve problems on the cellular level. Manipulation of the genetic information inside the body in real-time is only one example.

The principles of biological recognition could also serve as the basis for applications that seem to be very far from living systems as we identify them. One of the most obvious directions in the field of nanotechnology will be molecular electronics. The use of the tools of molecular recognition between biomolecules and their ability to self-assemble into elaborated structures could actually serve as an excellent model system for the 21st century nano-engineers. This course of study will link us directly to the one of the most important areas of research and development of the second half of the 20th century, the silicone-based microelectronics. Biology-based nanotechnology may facilitate our ability to overcome many of the limitations of the silicon-based world as we know them today. The field of electronics could be transformed to have complete new and exciting prospects when electronic devices will be made by nano-machines and by govern by molecular self-assembly principles. The world after this bionanotechnologic revolution may be a place where the molecular mechanism of thinking and reasoning will be understood and truly artificial intelligence machines could be manufactured.

It should always be remembered that the new era of nanobiotechnology and bionanotechnology may also lead to a situation in which there will be a very fine line between prosperity and devastation. The role of the scientists and technologists in the years to come will be to ensure the proper usage of these hardly imaginable abilities. We must make sure this revolution will be used for the benefit of mankind and not for its destruction. As those tools are so powerful, their misuse may actually lead to grave consequences.

E. Gazit

Chapter 1

Introduction: Nanobiotechnology and Bionanotechnology

The convergence between the scientific disciplines of biotechnology and nanotechnology is a relatively recent one. Yet, the combination of these highly important areas of research has already resulted in remarkable achievements. This chapter will provide basic background on classical and modern biotechnology and its interface with nanotechnology, a much less mature scientific discipline. Throughout this book we will use the term "nanobiotechnology" to describe the applications of nanotechnology techniques for the development and improvement of biotechnological process and products. This includes the use of fabrication and manipulation techniques at the nano-scale to form more sensitive and accurate diagnosis methods such as a "lab-on-chip" and real-time nanosensors. This also includes the use of matrixes with nano-scale order for the controlled release of drug as well as tissue engineering and regeneration. Later on in the book, we will describe more advance futuristic directions such the development of manipulation devices at the nano-scale ("nano-robots") to perform medical procedures in patients or nano-machines that will serve as an artificial alternative to functional biological organs.

The term "bionanotechnology" will be employed to describe the use of biological building blocks and the utilization of biological specificity and activity for the development of modern technology at the nano-scale. Such bionanotechnology practice is obviously not limited to biological applications but have much wider scope. For instance, future applications of bionanotechnology could include the use of DNA oligomers, peptide nanotubes, or protein fibrils for the fabrication of metal nanowires,

1

interconnects or other physical elements at the nano-scale, that could be used in molecular electronics and nano-electromechanically applications and devices (Braun *et al.*, 1998; Reches and Gazit, 2003; Zhang, 2003; Patolsky *et al.*, 2004a,b).

1.1. Classical Biotechnology: Industrial Production Using Biological Systems

Biotechnology is a well-established scientific discipline. The practice of classical biotechnology, defined by the American Heritage Dictionary as "The use of microorganisms, such as bacteria or yeasts, or biological substances, such as enzymes, to perform specific industrial or manufacturing processes", goes back to the first half of the 20th century (Figure 1.1). The large-scale industrial application of biological processes, such as the production of acetone by the fermentation of starch with the bacterium *Clostridium acetobutylicum*, was already performed in 1916. The industrial production of penicillin antibiotic from *Penicillium notatum* bacteria for practical medical use was already achieved in the early 1940s.

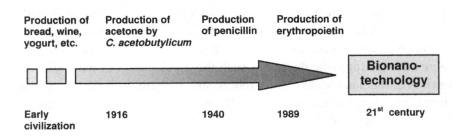

Figure 1.1: Milestones in the development of modern biotechnology into bionanotechnology. Production of bread, wine, and yogurt was practiced thousands of years before the commercial production of acetone, antibiotics, and biotechnological drugs.

Other scholars attribute the use of biotechnology practice to a much earlier era as a variety of food products, such as wine, beer, cheese and bread had already been produced with the help of yeast and bacteria for few thousands of years from the early days of civilization (Figure 1.1).

1.2. Modern Biotechnology: From Industrial Processes to Novel Therapeutics

Over the years, the practical definition of biotechnology became much more vague. Much of the activity of the modern biotechnology practice, both in the academic settings as well as the industrial ones, developed into applied biological sciences rather than a well-defined scientific discipline. The activity that is descried as "biotechnology" ranges from the production of biomolecules (e.g. functional proteins or antibodies) to serve as drugs to the development of novel diagnostic tools that are based on the interaction of specific biomolecules (such as antibody-antigen interactions as in immunodiagnostic kits or complementary nucleic acids interactions as in DNA-array microchips). The somewhat practical distinction between the pharmaceutical industry and the modern biotechnology industry is the production of large biomolecules such as functional proteins and antibodies by the latter as compared to small molecule drugs by the former.

For example, the first product of the largest biotechnology company, *Amgen*, is an industrially produced naturally occurring protein called erythropoietin (commercial name: EPOGEN®) that stimulates the production of red blood cells. This product was approved for human use in 1989 by the Food and Drug Administration (FDA) and represents one of the very first modern biotechnology products. Other biotechnology products include recombinant human insulin, human interferon, human and bovine growth hormones, and therapeutic antibodies. The production of therapeutic antibodies is especially intriguing and represents a relatively new area of research. These therapeutic agents utilize the remarkably affinity and specificity of antibodies (as described in Chapter 4) to specifically block harmful biological processes by a directed and extremely specific manner. A recent example is the humanized recombinant monoclonal antibody that binds to and inhibits the biochemical activity of the vascular endothelial growth factor (commercial name: AVASTIN™). This antibody-based treatment significantly increases the survival rates of patients with metastatic carcinoma of the colon that is incurable by conventional chemotherapy. Another major biotechnological product that was mentioned above is human interferon, a protein that is a central role in the response of the human body to viral infection, and human growth hormone, a key regulator of normal growth of children and adolescents.

Many more proteins are now been developed for their use as potential drug. One of the limitation factors for a much more common use of protein and peptide drug is their unavailability in oral formulation. Unlike small molecule drugs, which are commonly being administrated in the form of tablets or syrup, protein and peptide drug are usually being degraded during their passage in the digestive track. Therefore, the administration of the protein and peptide drugs as mentioned above is limited to injections that can not easily be performed by the patients outside of medical institutions. Another application of peptide and protein drug is topical, which again is very limited in its therapeutic scope. Actually, in this case nanotechnology can provide a very helpful and important solution. For example, arrays of hundreds or thousands of tiny nano-syringes are used for dermal application of biomolecular agents without any associated pain. Other applications include various nano-carriers are being studied for their ability to safely transfer drugs throughout most of the digestive track, but release them at the intestines. Nano-carriers are also being studied for their ability to allow the transfer of protein and peptide drugs through the blood-brain barrier (BBB) to treat disease such as brain tumors and various neurodegenerative disorders. The nano-carriers themselves can be formed by the self-assembly of biomaterials (such as peptide nano-spheres) but can also be composed of non-biological materials. At any rate, this is an example for the way in which nanotechnology can help biotechnology to reach much wider use.

1.3. Modern Biotechnology: Immunological, Enzymatic, and Nucleic Acids-Based Technology

The diagnostic aspect of biotechnology principally consists of the detection and quantification of biological materials using biochemical techniques such immunological recognition assays, enzymatic reactions, and DNA- or RNA-based technology. As will be discussed in this book, the use of nano-scale assemblies and fabrication may help to significantly improve the sensitivity as well as the specificity of the diagnosis process.

Diagnostic immunoassays include products such as the home pregnancy test kits that contain antibodies that detect minute traces of the human chorionic gonadotropin (hCG) hormone (Cole, 1997). Other products include diagnostic kits for the determination of HIV or hepatitis virus infections. In all these cases a specific high affinity and specificity molecular recognition processes facilitate the diagnosis process. The chemical basis for this recognition, specificity and affinity will be further elaborated in this book. Understanding and miniaturization of the detectors and molecular markers of the recognition process will lead to better utilization of these interactions for various diagnostic applications. The increased sensitivity of such devices will result in the need for much lower volume of blood to determine specific key biological parameters such as the glucose levels (Figure 1.2). Actually this lower amount of blood, that could be reduced from micro-liters to nano-liters, can be extracted by nano-syringes that were mentioned above that will significantly reduce the discomfort that is associated with the collection of blood samples for glucose levels determination. A more futuristic instrument may include a nano-device that measures the glucose levels using electrochemical reaction and nano-electrodes on a chip that are connected to a controlled delivery system that releases insulin according to a programmed profile. Such a nano-device will be able to mimic some of the functions of the pancreas for acute Type I and chronic Type II diabetes patients.

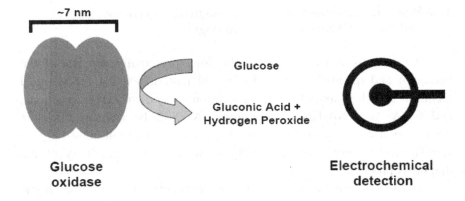

Figure 1.2: Detection of glucose level is based on the combination of enzymatic reaction and electrochemical detection. Hydrogen peroxide is being formed upon the oxidation of glucose by the glucose oxidase (GOX) enzyme. The level of the redox reaction is being determined electrochemically. In principle the reaction can be performed on the nano-scale as the size of the enzyme is less than 10 nm.

The diagnostic use of enzymatic reactions includes common instruments such as home glucometers that utilize the catalytic reaction of the glucose oxidase enzyme with glucose to produce gluconic acid and hydrogen peroxide (Jaffari and Turner, 1995). The later chemical product is being electrochemically determined and converted into a digital output (Figure 1.2). This is indeed one of the key examples of the merging between biotechnology and electronics to produce a hybrid enzyme-electronic interface. Miniaturization of such and other enzyme-based detectors is one of envisioned nanobiotechnological devices (Jaremko and Rorstad, 1998).

DNA-based technology includes the polymerase chain reaction (PCR) process that is being extensively used in forensic science to clearly determine the source of biological samples. Another technique that utilizes the specificity of nucleic acids interaction is the DNA microarray chip technology. This technology allows us to follow the expression pattern of thousands or even tens of thousands of genes simultaneously. The DNA-array techniques are not only invaluable to basic medical studies but it also paves the way for future personal medicine. Also these applications can be significantly advanced by the use of nanotechnology, for example by performing the reactions at very small volume using "lab-on-a-chip" settings. This should allow much smaller amounts of biological samples for the application of the nucleic acids based techniques. Such an increase in sensitivity should be especially helpful in forensic science and in the early detection of pathological processes such as early phase of malignant transformation. Other very important applications include environmental monitoring and homeland security detection of biological and chemical agents.

1.4. The Interface Between Nanotechnological and Biotechnology: Bionanotechnology

As mentioned above, nanotechnology is a relatively young scientific discipline. Chapter 2 provides some of the basic themes and principles of nanotechnology that are mainly directed towards readers with a background in the biological sciences. As a basic introduction, nanotechnology could be described as technological applications using systems and devices with orders at the nanometer scale (as most of the readers are aware, one nanometer equals to one thousandth of a micrometer or one millionth of a millimeter - 1/1,000,000,000 meter). The study of nanotechnology includes molecular systems, molecular assemblies (such as quantum dots), and organized self-assembled devices and machines. All these are considered to be part of the "bottom-up" approach in which simple components at the nanometric scale join together to form complex machines, devices, or instruments, by spontaneous processes that involve molecular recognition and self-assembly events.

Biology and biological recognition should have great importance in these directions of molecular recognition and self-assembly. Biomeolcules and biological complexes are naturally-occurring building blocks, recognition modules, and even machines (such as the ribosome, the complex protein assembly line, or molecular motor). Even more complex structures such as plant and animal viruses or bacteriophages (viruses that attack bacteria) could be formed by association at the nano-scale. Biological recognition, such as the one attained by specific antibodies (see Chapter 4), can specifically direct the "bottom up" assembly. By the utilization of the high specificity and the spontaneity of biological association processes, biological assemblies can serve as a "smart scaffold" for the self-association of complex organic and inorganic nano-machines and nano-devices.

Other directions are the "top-down" approaches which consist of either continuous improvement of common ultraviolet (UV) lithographic processes or the use of more elaborate lithographic technologies such as electron-beam (E-beam) lithography or focused ion beam (FIB) lithography. As of 2006, industrial scale lithography, used in the field of micro-electronics, already reached a 90 nm resolution by the use 193 nm deep ultraviolet photolithography. Electromagnetic radiation of shorter wavelength was already demonstrated to be useful at even better resolution (down to 45 nm). Furthermore, the use of beam of electrons or ions instead of photons allows the fabrication of objects at even higher resolution (as low as 20 nm).

There is not doubt that these techniques, either the "bottom-up" or "top-down", will revolutionized the field of biotechnology. The above mention biotechnological techniques and procedures and many others could be extensively benefited from their interfacing with modern nanotechnology techniques. This may be merely the miniaturization that will result in much simpler procedures that will require significantly smaller volumes of biological samples and will make the medical procedures much simpler and more pleasant to the patient. Yet, it could lead to assembly of sophisticated machines at the nano-scale that do not only measure but also perform various tasks that are based on these measurements. Moreover, the interface of biotechnology and nanotechnology can offer the attainment of sensitivities that are orders of magnitudes higher than current techniques. The ability to determine the presence of the very first cancerous cell in the body or the monitoring of minute amounts of pollutants or volatile explosives will dramatically revolutionize the medical practice as well as homeland security and environmental monitoring.

1.5. Supramolecular (Bio)Chemistry: The Theoretical Basis for Self-Assembly

Supramolecular chemistry consists of the study of the structure and function of the entities formed by association of two or more chemical species by non-covalent interactions. These type of supramolecular structure are actually type of "supermolecules" that are distinct from regular molecules that are being formed by continuous chemical covalent bonding between atoms. To distinguish between supramolecular structures and covalent chemical structures, we will usually use the term "assemblies" to describe the non-covalent structures.

The foundations of this field of science were laid down only few decades ago. The founding fathers are considered to be Donald J. Cram, Jean-Marie Lehn, and Charles J. Pederson who were awarded the Nobel Prize in Chemistry in 1987. The citation of the prize reads "for their development and use of molecules with structure-specific interactions of high selectivity". According to one definition suggested by Lehn: "supramolecular chemistry is chemistry beyond the molecule". It should be noted that the term "supermolecules" was already coined in the 1930s (Pfeffer, 1927), yet the exact description of the complex chemical interactions that lead to the formation of these ordered structures was yet to be described decades later.

It is also worth to note that while this notion of supramolecular assemblies is relatively new, many biological entities can consider as supramolecular assemblies. These can be as simple as functional dimers (e.g., Fos-Jun transcription factor) and as complex as molecular nano-machines such as the ribosome or molecular motors. Taken together, biology can be considered as a most advanced playground of supramolecular chemistry as complex biology functions is being achieved by the formation of complex molecular structures by non-covalent interactions (Aizenberg *et al.*, 2005). Throughout this book we will consider self-assembly and molecular-recognition (see below) as central themes in bionanotechnology and nanobiotechnology.

These processes in which large ordered assemblies are being formed by non-covalent association of simpler building blocks are the essence of "bottom-up" design. In this approach simple building blocks are joined together to form supermolecules or assemblies that have a distinct morphology and often also specific function or unique physicochemical properties.

1.6. The Next Steps for Self-Association at the Nano-Scale

The formation of ordered structures by self-association at the nano-scale is just the first steps toward the formation of ordered molecular structures. The next steps involve the formation of hetero-component assemblies, such as observed with the biomolecular assemblies and machines described above. The formation of such structures should be based on the combination of recognition modules that are evolve from geometrical complementary (Figure 1.3) but also on chemical recognition. Elements such as aromatic moieties that combine both properties are especially attractive elements in the self-association design schemes. This is due to the geometrically restricted nature of these interactions on the one hand, but the specific entropic and enthalpic chemical interaction on the other hand. It has already been shown that well-ordered structures, such as peptide nanotubes could self-assemble by the facilitation of such interactions (e.g., Reches and Gazit, 2003).

As will be discussed in Chapter 4, it was already demonstrated that many complex structures could be formed by the self-association of biomolecules. The pioneering work by Nadrian Seeman and co-workers had demonstrated the ability of partially complementary DNA strands, which form molecular junctions, to self-assemble into two-dimensional nanowires and three-dimensional nanocubes. Other researchers had demonstrated the ability of peptide molecules to self-assemble into tubes, spheres, plates, tapes, and hydrogels with nano-scale order (Ghadiri *et al.*, 1993, 1994; Agelli *et al.*, 1997; Altman *et al.*, 2000; Bong *et al.*, 2001; Banerjee *et al.*, 2003; Silva *et al.*, 2004). Such nano-scale peptide assemblies were already proven to have potential in diverse fields such as tissue engineering and regeneration (Holmes *et al.*, 2000; Silva *et al.*, 2004), the design of novel antibacterial agents that form nanochannels (Ghadiri *et al.*, 1994), and the fabrication of metallic nanowires (Reches and Gazit, 2003).

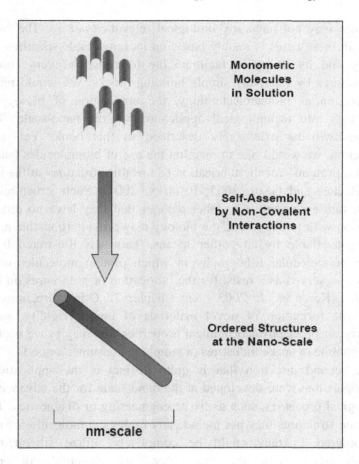

Figure 1.3: Self-assembly of molecules into ordered non-covalent structures as the nano-scale. In the current example complementary geometry is directing the association. Yet, other factors such as chemical complementary may facilitate such multi-component association.

1.7. Biology in Nanotechnology and Nano-Sciences in Biotechnology

As already mentioned above, the practice of biology or biotechnology at the nano-scale could be divided into two main categories (Figure 1.4). The first one is the technological applications of self-assembled nano-scale ordered biomolecular in various fields. These disciplines may not have any biological relevance *per se*. The role of biology in these cases is mainly based on its remarkable specificity and diversity and its ability to facilitate the formation of very complex nanostructures by relatively simple building blocks. We would refer to this direction as bionanotechnology, the application of biology and bimolecules into technological applications at the nano-scale. These directions will be extensively described in this book. Yet, as an introduction, we would like to mention the use of biomolecules (such as DNA and proteins) for the fabrication of metallic structures at the nano-scale (Reches and Gazit, 2003; Patolsky, 2004). Such structures can serve in future computers or other devices that may have no direction association with biology. Yet, the biology may provide tools that are not currently available by any other means. This was illustrated by the practice of molecular lithography in which protein molecules of few nano-meters served as a resist for the fabrication of gold wires on DNA molecules (Keren *et al.*, 2003 - see Chapter 7). Other directions may include the formation of novel materials of unique rigidity, surface chemistry, or other physicochemical properties that may prove useful in the automobile or space industries or simple in consumer's goods.

The second direction that is quite distinct is the application of technologies that were developed at the nano-scale for the advancement of biological processes, such as tissue engineering or diagnostics. These nano-scale structures may not include any biological molecules of nano-scale ordered as they could be completely utilize silicone-based structures (such as in the case of a "lab-on-a-chip") or carbon nanostructures (such as in the case of tissue engineering on carbon nanotubes-based substrate). Such an application could actually revolutionize the field of biotechnology by allowing the early diagnosis of disease, the on-line monitoring of therapeutic procedures, and the formation of tissue in the test tubes.

It is worth mentioning that the actual use of either the terms "nanobiotechnology" or "bionanotechnology" is quite a recent one. According to the *MedLine* database, the term "nanobiotechnology" appears first in 2000, while the first hit for "bionanotechnology" is only in 2004. "Nanobiotechnology" is also a much more common term that is more than 4 times more common that "bionanotechlogy". Yet, in many of the cases the use of the later term would have been much more appropriate.

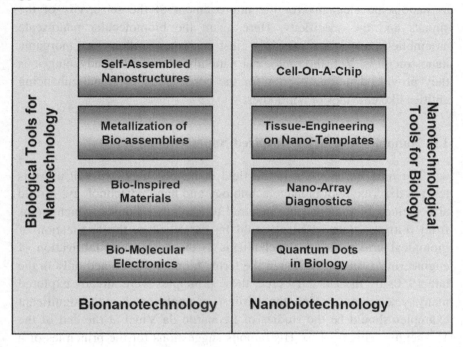

Figure 1.4: Bionanotechnology as the use of biological assemblies for various applications that may not be directionally associated with biology. Nanobiotechnology is the use of nano-science for specific biological applications.

1.8. The Combination of Bionanotechnology and Nanobiotechnology

In spite of the distinction between bionanotechnology and nanobiotechnology, it should be stressed for future applications, especially complex ones that both fields could be merged. For example, in the field of tissue engineering, self-assembled biomolecular structures could serve as a three-dimensional template for the formation of complex organs in the test-tube. In these applications, the biomolecules serve exactly as nano-carbon structures as mentioned above. Nevertheless, the biomolecular nano-scaffold may provide unique advantages such as biocompatibility and the ability for decoration with biological signaling motifs. In the field of nano-bio-diagnosis peptide nanotubes could serve as nano-scale elements for the improvement of the sensitivity of the signals and the specificity. Here again the biomolecular nano-scale assemblies serve in principle just as the carbon or inorganic nanostructures. Yet, the biological nanotubes have many advantages as they provide a unique anchor for the conjugation of signal-enhancing entities like enzymes or antibodies.

1.9. Nanobionics and Bio-Inspired Nanotechnology

The final introduction to the field of nanobiology (we will use this term to describe both bionanotechnology and nanobiotechnology) should also include bionics and bio-inspired technology. Bionics, which stems from [bi(o)- + (electr)onics] could be described as the application of biological principles and mechanisms to the design and fabrication of engineering systems. Although the term "bionic" was coined only in the late 1950s by the US Air Force, these principles were already explored many years before. One the earliest examples and most significant examples should be the studies of Leonardo da Vinci at the end of the 15th century (Figure 1.5). His famous suggestions for the principles of a helicopter were based on his studies and extensive observation of the flight of bird. Other bionic applications include the use of the nose of the dolphin as a model for a pear-shaped bow protuberance of ships or self-cleaning surfaces that are based on the surface properties of lotus leaves. It should be stressed that the aim of bionics is not merely to copy nature but rather to understand its principles and use them as a stimulus and motivation for innovations. Thus, bio-inspired technology may be more suitable name to describe this activity.

As most of the bionics is based on the formation of macroscopic objects and machines, it could also include nano-scale machines. It is expected that in years to come, the same mimicking as was preformed in the macroscopic world will also be applied to the microscopic and nano-scopic world. This includes simple applications such as self-cleaning surfaces at the nano-scale but could be explored for the formation of much more complex structures such as molecular motors or even molecular machines.

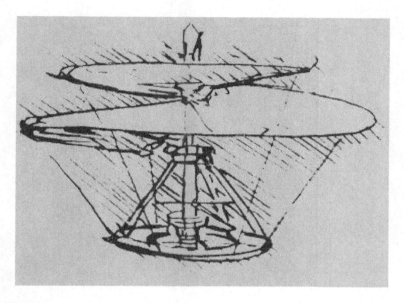

Figure 1.5: The drawing of a flying machine, based natural principles of the on flying mechanism of birds, made by Leonardo da Vinci at the end of the 15th century.

Chapter 2

A Brief Introduction to Nanotechnology

Nanotechnology, in its broadest definition, is related to the study and technological applications of materials and assemblies at the nanometric scale. The common symbol for nanometer is "nm" (as a reminder, one nm is one billionth of a meter - about the size of a small, molecule that consists of few atoms). The prefix *nano* originated from the Greek word *nanos* which literally means "a dwarf". Traditionally nanotechnology was considered to be related to the study of objects smaller than 100 nm. However, for the discussion of nanobiotechnology and bionanotechnology at large, we will also discuss assemblies at the 100-500 nm range (e.g., small unilamellar lipid vesicles or liposomes that are being used for drug delivery applications). In this chapter we will give a brief summary of the history and current trends in nanotechnology research and applications without the assumption of any previous knowledge. This is done in order to give the appropriate introduction for readers with biological or other non-physical background. While the introduction is quite short, it should allow for the understanding of the basics of biological branches of nanotechnology, nanobiotechnology or bionanotechnology.

2.1. The Emergence of Nanotechnology: "There's Plenty of Room at the Bottom"

Most scholars attribute the inauguration of modern nanotechnology to a very famous talk given by Prof. Richard P. Feynman (Figure 2.1) on December 29th, 1959, at the annual meeting of the American Physical Society that was held at the California Institute of Technology. The legendary address, entitled *"There's Plenty of Room at the Bottom. An Invitation to Enter a New Field of Physics"* (Appendix I), provided the first methodical introduction for the establishment of a new field of molecular manipulation. Feynman, a true maverick and one of the leading physicists of the 20th century, who was awarded the 1965 Nobel Prize in Physics for his seminal work on quantum electrodynamics, suggested quite bluntly in his talk that it should be possible to manipulate molecules and atoms directly. Although he did not mention the term nanotechnology - his original statement was: "What I want to talk about is the problem of manipulating and controlling things on a small scale" - He was genuinely the first person to define the uniqueness and the enormous potential of studies at the nanometric scale. He clearly stated that the principles of physics do not deny the possibility of manipulating things atom by atom. In his exciting and stimulating address Feynman discussed many issue ranging from data storage and fast computers to lubrication and evaporation at very small scales.

The relevance of Feynman's talk, almost half a century after it was delivered, is quite astonishing. This of course reflects the vision and imagination of this great scientist. Many of the ideas that were raised in this classic public address are very relevant in our current time and even for future endeavors in nanotechnology. Actually some of the challenges that were put forward tens of years ago, such as the fabrication of nano-scale motors, have not yet been achieved.

While Feynman is best known for his seminal contribution to modern physics, a very interesting point relates to Feynman's great interest in and affection towards the study of biology. He rejects the patronizing attitude of some of the physicists in the 50s towards biologists - "I don't know any field where they are making more rapid progress than they are in biology today" - and suggests that what would eventually be termed as nanotechnology has both a lot to learn from biology as well as providing new and indispensable experimental tools for the biologists.

It will indeed take many years before biology will be integrated into the field of nanotechnology, although as was already mentioned and will be discussed below biology is the only realm in which handful of genuine nano-machines exist even in the most simple bacteria. Molecular nano-motors that we expect to be fabricated by advanced nanotechnology techniques are being produced by very primitive bacteria to serve as flagella and other swimming devices.

Figure 2.1: Richard P. Feynman (1918-1988). A true maverick and the founding father of modern nanotechnology. "An honest man, the outstanding intuitionist of our age, and a prime example of what may lie in store for anyone who dares to follow the beat of a different drum" - Dr. Julian Schwinger. Picture is reproduced with permission from the Nobel Foundation. ©The Nobel Foundation.

2.2. Coining the Term "Nanotechnology" and Emergence of the Nanotechnology Concept

The actual coining of the term *"Nanotechnology"* is attributed to Norio Taniguchi from Tokyo Science University, who first mentioned it in 1974 (Taniguchi, 1974). His definition for nanotechnology, which is still relevant in the 21st century, was: "'Nano-technology' mainly consists of the processing of separation, consolidation, and deformation of materials by one atom or one molecule" (Taniguchi, 1974).

Much of the later driving force to make nanotechnology a major scientific field should be credited to K. Eric Drexler (Drexler, 1981, 1986, 1992). In 1981 he had already published an article in the *Proceedings of the US National Academy of Science* in which he described how biological protein assemblies can serve as functional units such a motors, cables, and pumps at the nano-scale (Drexler, 1981). In this article Drexler very nicely compared the components that are available in every living cell are comparable to components that are used for mechanical engineering. Drexler's 1986 book entitled *"Engines of Creation: The Coming Era of Nanotechnology"* is a true milestone and a pivotal reference for the field on nanotechnology. This book is freely available on the internet (see the bibliography sections) and readers are highly encouraged to download it. A later, more technical yet excellent description of the foundations of nanotechnology and its potential uses could be found in Drexler's 1992 book *"Nanosystems: Molecular Machinery, Manufacturing, and Computation"*.

2.3. Manipulating Molecules: The Scanning Probe Microscopes

In his talk Feynman envisioned the ability to make smaller and smaller tools by a sequential miniaturization of machine shop tools - making tools that are one tenth the scale of the current tools, then use these tools to make new set of tools that are one hundredth scale of the common tools, and so on and so forth until the ability to manipulate atoms and molecules is achieved.

It took some years until such tools became available. Yet these tools came from quite a different origin. The major milestone was invention of the Scanning Tunneling Microscope (STM) by Gerd Binnig and Heinrich Rohrer at IBM's Zurich Research Labs (Binning *et al.*, 1982), who won the 1986 Nobel Prize in Physics for this achievement. This was followed five years later by the invention of the Atomic Force Microscope (AFM). This type of microscope is quite different from any other microscopes, either light or electron microscopes. Unlike conventional microscopes in which a beam of photons or electrons are used for the construction of images that are based on optical principles using optical or electromagnetic lenses, the STM and AFM microscope are based on very fine scanning of the sample surface using an ultrafine tip. The image that is achieved is based either on the electrical tunneling properties of the matter (in the case of STM) or the topographical properties of the sample (in the case of AFM). The generic name for this type of microscope is known as scanning probe microscope (SPM). The tip, known as probe, scans over the surface of the sample and measures tunneling current in the case of STM or the physical topography of the surface in the case of AFM (Figure 2.2). Older readers may think about the scanning of the tip as the diamond stylus that rides along in the groove of a vinyl record played on an old gramophone.

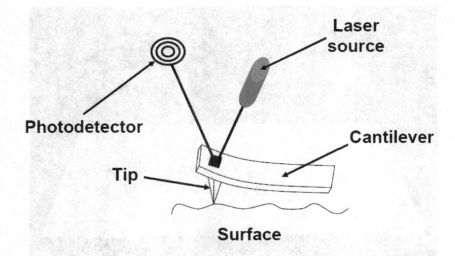

Figure 2.2: The principle of action of the atomic force microscope (AFM). A very fine tip travels over the surface of the sample. The topographic properties of the scanned surface affect the reflection of the laser light from the cantilever. The changes in the reelection are being translated into a three dimensional map of the scanned surface.

In principle the specifications of the scanning probe machinery allow us to reach a single atom resolution and indeed the atomic organization of metallic substrates. In the case of biological samples STM generally could not be used as conductivity of the sample is required for tunneling current, yet most of the biological molecules are insulators. Also in the case of AFM, atomic resolution could not be reached with biological samples due to the soft nature and stickiness of the biological samples that result in interaction with the tip and stretching of the sample in the low nanometric scale. A significant improvement in the study of biological and other "soft materials" came with the introduction of tapping-mode AFM. In this mode, the probe oscillates at a resonant frequency and at amplitude setpoint while scanning across the surface of a sample. In that way, the amplitude of the oscillation of the probe is changed in relation to the topography of the surface of the sample and thus it allows the reconstruction a topographic image.

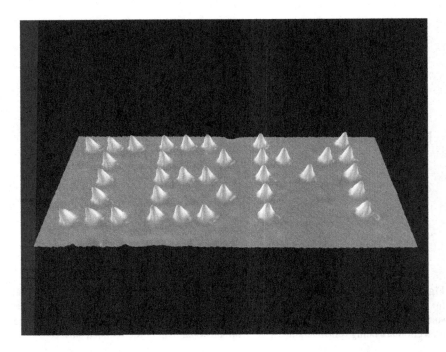

Figure 2.3: Spelling of "IBM" using 35 xenon atoms on a nickel surface. Courtesy of Dr. Donald Eigler. Image reproduced by permission of IBM Research. Unauthorized use not permitted.

The first vivid demonstration of the ability to manipulate atoms at the nano-scale was the ability to spell "IBM" using 35 xenon atoms on a nickel surface, using a STM to move the atoms into their position (Figure 2.3; reproduced from Eigler and Schweizer, 1990). This demonstration by the IBM Fellows was later followed by the construction of the famous "quantum corrals" in which iron atoms are positioned in circular patterns. These atomic constructions are made by bringing the STM tip very close to the atoms in a way that the weak interatomic and chemical forces between the tip and an individual atom move the atom to the desired position.

Scanning probe microscopy was later used for much extended molecular organization purposes. The Dip-Pen Nanolithography (DPN) technique that was developed by Chad A. Mirkin and coworkers uses AFM to introduce patterned molecules into surfaces (Lee *et al.*, 2002, 2003). In this technique the AFM tip is used to deliver molecules to a surface via a solvent meniscus, which naturally forms at the end of the tip in ambient atmosphere. The molecules are being used as an "ink" while the AFM tip is being used as a "pen". This technique is very useful for the patterning of biological molecules at a resolution that is better than 100 nm.

2.4. Carbon Fullerene: A New Form of Carbon

Much of the current everyday chemistry is based on the use of organic compounds, which means molecules that are based on carbon as a key building block. Almost one fifth of the weight of the human body is carbon, which is more than any other element except oxygen. Not only most biological entities such as proteins, nucleic acids, and lipids are organic compounds but much of the common industrial products such as nylon, plastic, Teflon, *etc.* are made of carbon as a key constituent. Also new materials including hyper-strong but light-weight materials such as KEVLAR® are based on organic entities such as aromatic amides. The use of aromatic moieties for fibril formation will be described later in the book. Organic chemistry in general is a very advanced field of chemistry and the synthetic abilities in this field both in terms of complexity of the synthesis as well the scale-up of the processes are enormous.

Carbon can be obtained in nature in several forms. Until 1985, the known allotropic forms of carbon were amorphous, graphite, and diamond. This by itself shed some light of the diversity of carbon organization at the molecular level, as graphite is one of the softest materials known in nature while diamond is of course one of the hardest. That implies that the precise molecular organization of the carbon can dramatically affect its macroscopic properties.

Figure 2.4: The new form of C_{60} carbon clusters. The closed caged buckminsterfullerene are being formed by the arrangement of pentagons and hexagons into soccer-shaped closed-cage.

A new form of carbon was identified in 1985 by Curl, Smalley, and Kroto who shared the 1996 Nobel Prize in Chemistry for this discovery (Kroto *et al.*, 1985). During experiments in which graphite was vaporized using a laser, new forms of carbon clusters were discovered. These clusters were in the form of fullerenes, closed and convex cage molecules formed by the arrangement hexagonal and pentagonal faces (Figure 2.4). The most common and most stable carbon cluster was the buckminsterfullerene, named after the architect R. Buckminster Fuller due to the resemblance of the carbon structure between Fuller's geodesic domes. This caged cluster is a 60 carbon atoms spheroidal molecule, resembling a soccer ball in which hexagon carbons are arranged to form ordered closed structure pentagon carbon faces that serve as vertexes. Therefore, another popular name for the buckminsterfullerene is buckyball. This is in contrast to graphite which is composed of a two-dimensional flat layer of hexagonal carbon arrangements. The C_{60} Buckminsterfullerene is the most abundant cluster of carbon atoms found in carbon soot. Larger carbon fullerenes that have 70, 76, 84, 90, and 96 carbon atoms were also found to be formed under similar experimental conditions.

The size of a C_{60} fullerene is about 0.7 nm and it shows excellent electrical and heat conductivity. Even smaller closed-cage carbon cluster are possible to obtain including C_{50}, C_{36}, and C_{20}. The smallest possible fullerene, C_{20}, consisting solely of 12 pentagons, has been generated from a brominated hydrocarbon, dodecahedrane, by gas-phase debromination. Yet, the lifetime of the C_{20} was only about 0.4 milliseconds. It may be possible that in the future ways will be found to stabilize these sub-nm structures. On the other hand much larger carbon clusters are also found. Such closed-cage structures contain hundreds of carbon atoms. There were also descriptions of "carbon onions" in which for example a C_{20} was placed inside a C_{240} cage which was in turn inside a C_{540} cluster cage.

Many applications have been suggested for the carbon fullerenes. Many of the biological or biotechnological ones are related to the ability of the assembly to serve as nano-containers for drug delivery or a reservoir to diagnostic agents. It was also demonstrated that modified fullerenes could serve as anti-viral agents. It was revealed that a modified C_{60} molecule could effectively inactivate human immuno-deficiency virus (HIV) at low micromolar concentrations (Schinazi *et al.*, 1993).

2.5. Carbon Nanotubes: Key Building Blocks for Future Nanotechnological Applications

The discovery of the closed-cage carbon structures was followed by the discovery in 1991, by Sumio Iijima, of yet a new form of carbon at the nano-scale, the carbon nanotubes. These nanotubes, in their single-shell form (known as single-walled nanotubes - SWNTs) are elongated variants of the C_{60} fullerens that described above. They practically resemble graphite sheets wrapped around into elongated cylinders with or without capping at their ends (Figure 2.5).

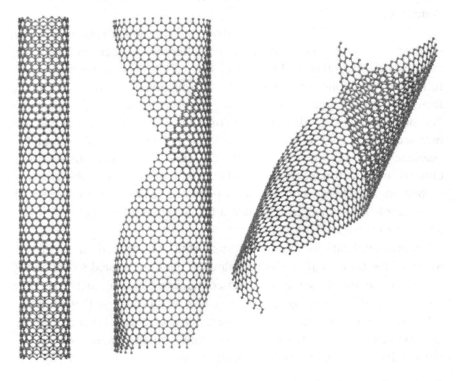

Figure 2.5: Model for the formation of carbon nanotubes by the two dimensional graphite-like layer. The tubes presented here are single-walled, although double-walled and multi-walled carbon nanotubes also present. Courtesy of Chris Ewels (©Chris Ewels www.ewels.com).

The formed nanotubes could be remarkably long. One key property of the carbon nanotubes is their remarkable aspect ration: their length to width ratio is extremely high (few nm in diameter and up to 1 mm in length). Carbon nanotubes could also be formed in the multiwalled fashion (know as multi-walled nanotubes - MWNTs). In this arrangement the tubes are concentrically nested like rings of a tree trunk (Baughman *et al.*, 2002).

Carbon nanotubes have unique physical properties: both mechanical and electrical. They have a tensile strength that is about 10 times greater than steel, yet at about one quarter the weight. Therefore, carbon nanotubes are considered the strongest known material for their weight. Therefore carbon nanotubes are expected to be used in the development of ultrastrong materials for the construction, car, and aviation industries. Moreover, carbon nanotubes are envisioned as thread for ultra-strong fabrics and clothing such as light weight bullet proof fabrics. One of the wildest ideas for the use of nanotubes is the assembly of an ultra-strong cable to be used for the construction of a space elevator.

Unlike a single sheet of graphite which is a semiconductor, carbon nanotubes are found as a random mixture of conductive and semiconductive entities. The conductivity of the tubes depends on the direction by which the sheet of carbon is wrapped to form the tube. The conductivity of tubes on the one hand and their small size on the other led to the suggestion of the use of the tubes as molecular wires, conductive composites, and interconnects. Furthermore, the electric properties of carbon nanotubes appear to be modulated by various gases, for example nitrogen dioxide. Thus arrays of carbon nanotubes could serve as very sensitive gas sensors.

Other potential applications that have been proposed for carbon nanotubes includes energy storage and energy conversion devices; field emission displays and radiation sources; hydrogen storage media; and nanometer-sized semiconductor devices (Baughman *et al.*, 2002). It is worth mentioning that many of the potential commercial uses of carbon nanotubes are limited by its high cost of production. A pure sample of carbon nanotubes cost, as of 2006, a few hundreds dollars a gram. Such a high price makes many of the suggested uses of carbon nanotubes commercially unrealistic. Yet, much efforts are being invested in the construction of industrial settings for large scale production of carbon nanotubes and their price is expected to be lowered in years to come.

2.6. Non-Carbon Nanotubes and Fullerene-Like Material: The Inorganic Nanomaterials

As discussed in the previous section, carbon nanotubes are made of two dimensional sheets of graphite that are rolled into a cylinder at the nano-scale. It is actually well-known that not only graphite, but also inorganic materials such as WS_2, MoS_2, NbS_2 can form two-dimensional layered arrangements (Tenne, 2002). Therefore in principle those structures could also be rolled to form either fullerene-like structures or nanotubes.

Figure 2.6: TEM micrograph of MoS2 fullerene-like nanoparticles. Courtesy of Reshef Tenne.

Indeed it was shown in 1992 by Reshef Tenne and coworkers that nanoparticles of the layered WS_2 compound can form closed-cage structures (Figure 2.6) that are very similar in their molecular organization to carbon fullerenes and carbon nanotubes (Margulis *et al.*, 1993). This instability was attributed to the highly reactive dangling bonds of both sulfur and tungsten atoms, which appear at the periphery of the nanoparticles.

This seminal work was followed by the demonstration of a variety of inorganic layered compounds such as MoS_2, $NiCl_2$, NbS_2, Tl_2O to form fullerene-like nanoparticles and nanotubes (Tenne and Rao, 2004). It was therefore claimed that this nano-scale arrangement is generic to layered materials and numerous others. The production of inorganic nanomaterials has already become industrial and many technological applications were suggested for these materials. One the most intriguing use of these compounds is their application as solid lubricants. These fullerene-like inorganic nano-materials have remarkable lubrication properties which could significantly increase the lifespan and efficiency of motors and other power tools with moving parts. Indeed the application of these nano-particles as additives for motor oil is currently being test by major car manufacturers. Other applications which are more relevant to biological applications including their use as lubricants of medical and dental devices.

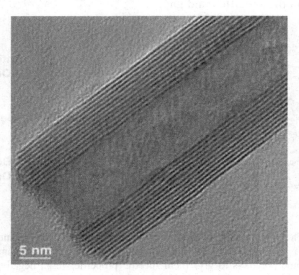

Figure 2.7: Transmission electron microscopy of WS_2 nanotubes. Courtesy of Reshef Tenne.

2.7. Quantum Dots and Other Nano-Particles

Other very intriguing assemblies at the low nanometric range are the quantum dots also known as nanocrystals (Empedocles and Bawendi, 1997). These nano-objects consist of semiconductor materials such as CdSe, CdTe, PbS, and PbSe surrounded by a shell that is usually composed of ZnS or CdS. These nanoassemblies have typical dimensions between a few nanometres to a few ten of nanometers (few tens of atoms - a 10 nm core is composed of about 50 atoms), but could also be larger. The quantum dots basically serve as a cage for electrons, and the molecular diameters of these structures reflect the number of electrons they contain. A typical quantum dot can have anything from a single electron to a collection of several thousands depending on their specific size. The quantum dot basically behaves like a giant atom. As in a regular atom, the energy levels in a given quantum dot become quantized due to the confinement of electrons (Figure 2.7). The band gap between the two energetic states of the giant atoms determines the energy of a photon that is being emitted by the transition between states. The energy of the photons can be either in the visible or the infrared region depending on the specific band gap.

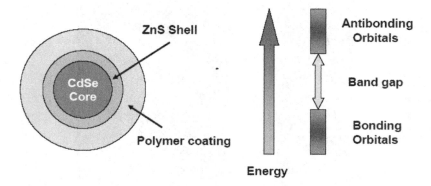

Figure 2.8: The organization of fluorescence quantum dots. The quantum dots are composed of a semiconductor core surrounded by a shell structures that are usually coated with some polymers to give them specific properties (e.g., negative or positive charge or functional chemical group). The fluorescence proprieties of the quantum dots arise from the energy gap of the confined electronic structures.

While quantum dots are extremely interesting form the physical point of view, for the purpose of this book, the most important property is their fluorescent nature. A given quantum dots exhibit high fluorescent brightness that is stable and long lasting, while having narrow emission span. The exact emission wavelength of a given tubes depends on its physical diameters. Commercially available quantum dots have emission wavelengths ranging from the viable 465 nm to the 2300 nm infrared region.

2.8. Nanowires, Nanorods, and Other Nanomaterials

In the previous part we discussed the formation of tubular and spherical and tubular nanostructures by carbon and inorganic oxides and sulphides. In all cases, layered materials wrap around to form cylindrical or spherical structures. In parallel to these works, other studies had indicated that many materials including metallic elements such as gold, silver or cupper as well as single element semiconductors (such as silicone or germanium), compound semiconductors (such as InP, CdS and GaAs as well as heterostructures), nitrides (such as GaN and Si_3N_4) to carbides (such as SiC) can form various ordered structures at the nano-scale. Such structures include small spherical nanoparticles, nanowires, nanobelts, nanoribbons, and nanorods. All these materials, especially the quasi-one-dimensional materials wires, ribbons, and rods, that have been attracting great research interest.

2.9. Magnetic Nanoparticles

The conductive and semiconductive nanomaterials attract much attention due to their potential use as in nano-scale electronics. Yet, of many other applications, most notable high-density data storage there is a need for magnetic particles. Indeed, different magnetic nanocrystals were prepared such from metal oxides as MnO, CoO, Fe_2O_3, $MnFe_2O_4$ and $CoFe_2O_4$. The size of the nanocrystals can be as small as five nanometers. Furthermore, recent studies have demonstrated the ability to form a single 30 nm crystalline magnetite (Fe_3O_4) nanotube (Liu *et al.*,

2005). Some research is focused on the coating of metallic magnetic nanoparticles with various organic materials. One example is the use of lipid coating with results in water-soluble magnetic nanoparticles. Magnetic nanoparticles are very useful both as the media for storage as well as part of nano-scale reading heads. Other very important application of magnetic nanoparticles is their use as contrast agents in magnetic nuclear imaging (MRI). This will be further discussed in Chapter 6.

Chapter 3

Natural Biological Assembly at the Nanometric Scale

The process of self-assembly is central to all biological systems. All living cells evolve from simple building blocks that spontaneously self-associate into complex networks, molecular machines, and cellular structures. This is true for all organisms, from the very simple bacteria that have only few hundred types of proteins for the building of their various nano-scale machines and assemblies to mammals that have tens of thousands of different and varied protein building blocks as part of their molecular arsenal. Sophisticated molecular machines, such as the ribosome, composed of tens of protein and RNA molecules that assemble into high-ordered and well-defined structures, is a key example of a hierarchal self-assembly of a nano-machine from rather simple building blocks. In this chapter we will provide some background to the process of self-assembly and self-organization in biological systems. We will describe supramolecular biological structures that have both physiological as well as pathological roles.

3.1. The Process of Self-Assembly and Self-Organization in Biology

Self-organization of supramolecular assemblies is the central theme of the biological systems (Whitesides *et al.*, 1991; Lehn, 2002). Unlike man-made engineered structures, complex three-dimensional biological systems are being spontaneously formed upon the association of simple building blocks into complex structures (Figure 3.1). The term "bottom-up", which was mentioned in the previous chapter as a future direction for nanotechnology, has been the natural choice for all living organisms.

The "top-down" approach is the usual tool for conventional micro techniques.

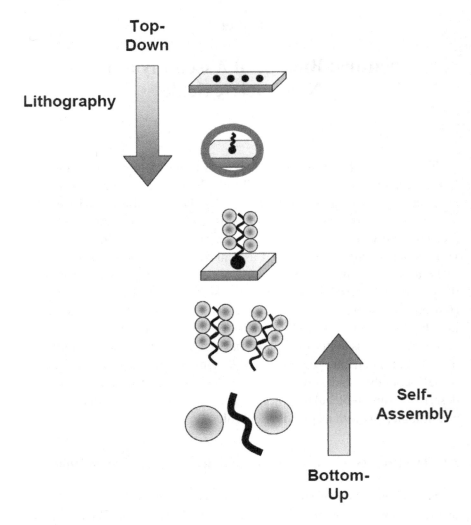

Figure 3.1: The hierarchal process of self-assembly into complex structures. The formation of order assemblies at the nano-scale can work either by the "bottom-up" organization of molecules to form an order structure or the engineering of a pre-formed pattern to serve as a template for molecular organization.

In most cases, the assembly of complex biological systems is not based on any pre-arranged templates, as we often see in engineering applications, but rather occurs through a process of molecular recognition and self-assembly events in which relatively simple building blocks recognize each other in solution, associate with one another, then associate with other molecules to finally form a well-organized three dimensional functional entity (Figure 3.1).

3.2. Organization of Bacterial S-Layers

One fundamental feature of carbon and inorganic nanomaterials, as discussed in Chapter 2, is the organization of these materials into well-ordered crystalline or nearly crystalline forms of one-dimensional, two-dimensional, and three-dimensional orders. Such organization of biological materials into ordered and crystalline nanomaterial could be also observed in several living systems. As will be demonstrated later in this book, these bioassemblies were already used in technological applications.

A key example, which was already suggested for nanotechnological applications many years ago, is the two dimensional bacterial S-layer assemblies. In a pioneering work, the structure, organization, and technological application of these assemblies were studied by Uwe B. Sleytr and co-workers (Sleytr *et al.*, 2005). S-layers are bacterial cell surface structures that are found in all archae and in hundreds of different species of walled bacteria (Figure 3.2). Archae, as may be known to some of the readers, is a very unique group of prokaryotes that includes some of extreme thermophilic and halophilic prokaryotes (DeLong and Pace, 2001). This interesting group of prokaryotes is phylogenetically distinct from bacteria and has some similarity to eukaryotes especially in their transcription and translation. Many of the organisms that belong to this group are extremophile (organisms that require extreme conditions) such as halophiles ("salt-loving" organisms that could survive solutions of several molars of salt such as in the dead-sea) and thermophiles ("heat-loving" organisms that could survive temperatures higher than 100 °C) The adaptation of archea and also some bacteria to such extreme environments may allow their application into stable nanotechnological devices for long-term usage, as the stability of common biomolecules is one of the limiting factors for their applications.

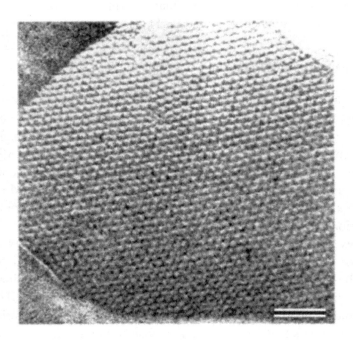

Figure 3.2: The S-layer structure. Transmission electron micrograph of a freeze-etched, metal-shadowed preparation of a bacterial cell with an S-layer with hexagonal lattice symmetry. The bar represents a distance of 100 nm. Courtesy of Uwe Sleytr.

As the assembly of S-layers have been optimized during billions of years of evolution, it is one the most well-organized and highly efficient biological self-assembling systems. These two-dimensional layers are composed of single protein or glycoprotein species that assembled into an ordered layer that completely covers the cell surface. This layer exhibits a clear lattice symmetry with spacing of the assembly units in the range of 3-30 nm. The thickness of the S-layers is at the nano-scale and ranges from 5-15 nm. They also possess pores of a seamless size and morphology in the 2-6 nm range. The very elegant use of these nano-scale ordered structures for nanotechnological applications will be described in Chapter 7.

3.3. Self-Organization of Viruses

Another well-studied biological process of self-assembly that was already suggested for nanotechnological applications is the spontaneous template-assisted formation of viral shells. Viruses and bacteriophages (virus that infect bacteria - from "bacteria" and Greek *phagein*, meaning "to eat") are very abundant in nature and represent the simplest form of autonomous biological systems. Although viruses obviously need the cellular machinery of the infected hosts for their replication they still represent an independent system that carries its own genetic information in a closed-caged environment. The defined molecular space that is isolated from the external world is a crucial element in self-replicating biological systems and represents the most fundamental aspect in the transition for a molecular collection into a living system. As will be discussed later in this chapter, it is usually achieved by the formation of lipid membranes for living organisms.

The most intensively investigated viral assembly is that of the tobacco mosaic virus (TMV; Figure 3.3). TMV was the first macromolecular structure to be shown to self-assemble *in vitro* in 1957. In a classic work Fraenkel-Conrat and Williams showed that upon incubation of the viral purified coat proteins and its RNA, virus particles formed spontaneously (Fraenkel-Conrat *et al.*, 1957). As will be described later for the ribosome, the specific interaction between the RNA nucleic acid and the proteins, results in the formation of well-ordered three dimensional structures. This observation suggested that all the molecular information needed about the precise assembly of the virus is retained with the building blocks. Furthermore, as the structure is formed spontaneously without the supply of any external energy, such as high energy nucleotides (such as ATP) as will be discussed for other-forms of self-assembly later in this chapter, clearly indicates that the self-assembled structure represent a minimal free energy supramolecular structure.

The first studies on the structure of TMV viruses were done in the 50s using electron microscopy techniques that were beginning to be established and revolutionized biological research in many senses. These particles are seamless, rigid, rod-like structures that form a hollow cylinder of about 300 nm in length, an inner diameter of 4 nm, and an outer diameter of 18 nm. Each viral particle contains more than 2,000 protein subunits that are arranged as a right-handed helix. The RNA molecule is retained between successive turns of the protein helix. The process of TMV self-assembly starts with the binding of a bilayer disk of protein building blocks to a site on the viral RNA scaffold. Upon binding of the RNA, the protein subunits of the disk undergo a coordinated rearrangement to yield a helical nucleoprotein complex.

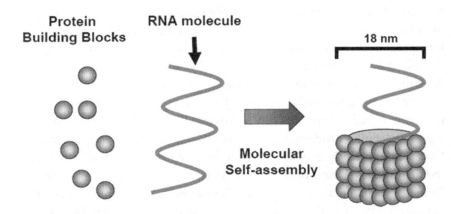

Figure 3.3: The formation of ordered TMV particles by the self-assembly of protein building blocks onto the RNA template. The RNA template serves as a pattern for the organization of the protein molecules to form a well-defined multi-component structure at the nano-scale.

As with the S-layers, the TMV viral nanoparticle is a very interesting template of bionanotechnology. As its three dimensional structure resembles many of those formed by inorganic or carbon nanostructures. Yet, unlike many of the pure carbon and inorganic nanostructures, these nanoparticles are formed spontaneously by self-assembly under mild conditions. Indeed, there were some reports on the metallization of TMV particles using electroless deposition (Kenz *et al.*, 2003, 2004). It is reasonable to speculate, that in the future, genetically modified protein units may be used for more precise TMV modifications for applications at the nanoscale.

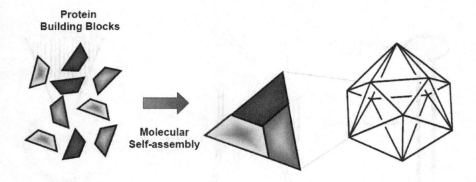

Figure 3.4: The formation of a icosahedral viral capsid by the self-assembly of protein building blocks into triangular elements that specifically interact to form the closed-caged assembly.

Another form of well-ordered viral assembly is formation of closed-caged structure that is composed of planar subunits that are closed together. This form of structure in known as icosahedral capsids, based on the symmetry of this platonic solid composed of identical subunits (Figure 3.4). Such caspids, e.g., the bacteriophage ϕX174, are built up by using three identical protein subunits to form each triangular face, thereby requiring 60 subunits to form a complete capsid. The readers should note the similarity between the organization of the icosahedral capsid composed of triangular faces and the quasi-icosahedral C_{60} carbon fullerene composed of hexagons and pentagons.

3.4. Self-Organization of Phospholipids Membranes

A self-assembly process that is crucial to any living organism is the spontaneous formation of lipid bilayers from phospholipid building blocks (Figure 3.5). Just as carbon and inorganic fullerenes, biological membranes are two dimensional layers that are closed in space to form a closed-cage spheroid structures. The main driving force of the formation of lipid membranes is hydrophobic interactions.

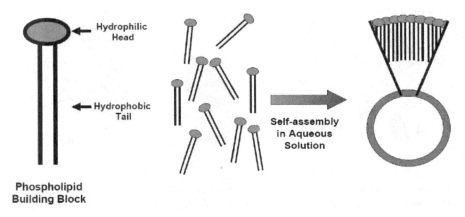

Figure 3.5: The formation of liposomes membranes by the self-assembly of phospholipids in aqueous solution. The driving force for the formation of the assemblies is the amphiphatic nature of the phospholipids building blocks.

The phospholipids building block is an amphiphilic molecule that has a hydrophilic part, the phosphate head group, and a hydrophobic, an aliphatic tail. Upon transfer of phospholipids molecules into aqueous solution a rapid and spontaneous association of the molecules occur during the enormous energetic cost of the presence of the aliphatic tail in water. The organization of the phospholipids is into bilayres structure in which each of the aliphatic parts of each layer are pointed towards each other and the hydrophilic heads are pointed either to the bulk water or to the inside part of the closed structure.

The similarity of these structures to those presented in Chapter 2 is not limited to the closed-cage organization. Upon drying of phosphoilipid from the hydrophobic organic solvent and its re-solvation in water, an "onion-like", larger multilammelar vesicle (LMV), of micron-scale are formed. Yet these "multi-walled" biological structures can be easily converted to small unilammelar vesicles (SUV) of single bilayers and sizes that range from few hundreds of nm to 50 nm upon sonication of extrusion. As other layered materials phospholipids can also form other organizations in the nano-scale the more common one is the hexagonal phase of phospholipid. Yet, tubular structures, similar to the nanotubes that were mentioned in Chapter 2 can be formed by the phospholipids under certain conditions (Figure 3.6). The ultrastructural analysis of these lipid nanotubes by electron nanotubes (Yager *et al.*, 1984; Yamada *et al.*, 1984; Nakashima *et al.*, 1995) reveals the similarity to the carbon and inorganic nanotubes as were described in Chapter 2. These are hollow structures with a sub-micron diameter that are assembled due the same physical principles that direct the formation of the bilayer structures. Thus it is free-energy driven spontaneous formation of well-ordered structures that is based on the closure of two-dimensional layer into a closed structure.

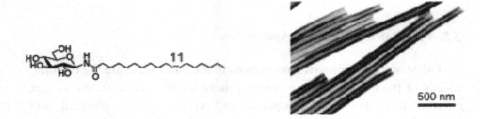

Figure 3.6: Formation of lipid nanotubes by the self-assembly of glucopyranosylamide lipids. The driving forces for the formation of the structures are as presented in Figure 3.6. Yet the specific properties of the hydrophilic headgroup modulate the final morphology of the formed assemblies. Reproduced with permission from Kamiya *et al.*, (2005). Courtesy of Mitsutoshi Masuda. Copyright (2005) American Chemical Society (ACS).

The formation of either tubular or spherical structures is based on the geometrical properties of the building blocks as well as the conditions of molecular assembly. The readers should note that the common characteristics of the formation of alternating spherical and nanotubular structures as were observed in carbon and inorganic nano-structures (Chapter 2), as well as the lipid nano-structures presented here, will be described for the aromatic peptide structures in Chapter 4.

Small lipid vesicles, also known as liposomes, are well-studied nanoparticles. Even before nanotechnology became a fashionable field of research, many studies have been conducted on the use of liposomes as vehicles for drug delivery. One of the best examples of the use of liposomes for drug delivery is the case of DOXIL®. This is a liposome injection of the anti-cancer doxorubicin drug that is successfully used to treat several forms of cancer with better efficacy and lower side-effects as compared to the parent drug.

There is a great interest in the use of decorated SUVs as targeted drug delivery agents. The basic premise is to conjugate some type of targeting devices, for example a receptor for a cancer cell, to a vesicle that contains chemotherapy agents. This should allow more effective treatment with much lower degree of side effects. Interestingly, liposomes are already used in the cosmetic industry where it is claimed that they help to transport agents such as vitamin E and other anti-oxidants upon the penetration of the vesicles into the inner parts of the human skin.

3.5. Fibrillar Cytoskeleton Assemblies

Other ordered biological assemblies in the nano-scale are the building blocks of the cytoskeleton, a complex network of protein filaments that extends throughout the cytoplasm and serves as the physical and mechanical support of the cell as well as intracellular cables for cargo movement. The cytoskeletal cellular main fibrillar elements are actin filaments, microtubules, and intermediate filaments which all are fibers in the nano-scale. These cytoskeletal elements facilitate the ability of cells to adopt a variety of three dimensional shapes and to carry out coordinated and directed movements. The mechanical rigidity of the cytoskeleton elements was determined at the nano-scale using AFM techniques (de Pablo *et al.*, 2003). By physical measurement of the force needed to deform these structures their stiffness (represented as Young's modulus, after the English physicist Thomas Young, the stiffness to withstand a physical load) was determined to be in the order of GPa which is high compared to other nano-scale ordered structures.

Actin filaments (also known as *microfilaments*): These filaments are ubiquitous to all eucaryotic cells and the building blocks of the filaments, the actin proteins, are well-conserved in evolution. Earlier analogues of actin, the FtsZ proteins, are found in bacteria, and also have structural roles in the prokaryotic organisms. These are two-stranded helical polymers with a diameter of about 7 nm (Figure 3.7). These are quite flexible structures that are organized into a variety of linear bundles. The actin filaments are the principal components of the thin filaments in skeletal muscles and they have a key role in cell motility.

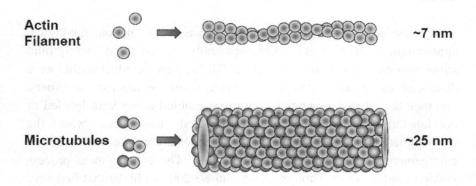

Figure 3.7: The major cytoskeleton nano-scale assemblies. The actin two-stranded helix is formed by the self-assembly of a monomeric actin building blocks and have a diameter of about 7 nm. The microtubules are being formed by the self-assembly of the tubulin dimeric building blocks to assemblies of about 25 nm in diameter.

Microtubules: These are hollow cylinders of out diameter of about 25 nm that are composed by polymerization of the tubulin protein building blocks (Figure 3.7). The microtubles are much more rigid as compared to the actin filaments. The microtubules network is attached to a central cellular entity known as the centrosome. Motor proteins, such as kinesin and dyneins move along the microtubules to transport all kinds of cargo within the cell. While kinesin moves only in an "outbound" direction away from the centrosome, the dyneins is the "inbound" train moving only inbound towards the center of the cell.

Intermediate Filaments: These filaments have ropelike cellular fibers with a diameter of about 10 nm. Their name comes from their intermediate dimension that is higher than the actin microfilaments and the much larger microtubules. Unlike the actin filaments and the microtubules that are each composed of a single protein building block, actin and tubulin respectively, these intermediate filaments are made of a large and heterogeneous family of proteins. These filaments have diverse roles such as mechanical support for the cell and the formation of a network around the nuclear membrane.

Cytoskeleton polymers were already used for nanotechnological application. Actin filaments were chemically reacted to covalently bind silver nanoparticles (Patolsky *et al.*, 2004). The modified filaments were disassembled, excess silver was removed from the labeled monomers, and then the labeled monomers were reassembled either with labeled or non-labeled co-monomers. These labeled monomers with the reassembled monomers were used as nucleation site for catalytic enlargement of the nanoparticles with gold. The enhancement process yielded gold wires (1-4 micron long and 80-200 nm high) that had very good electrical conductivity (Patolsky *et al.*, 2004). Furthermore, the ATP-fuelled motility of the actin-Au wire-actin filaments on a myosin interface was also demonstrated (Patolsky *et al.*, 2004). This provided a vivid demonstration for the use of biological templates for the fabrication of inorganic materials at the nano-scale.

3.6. Nucleic Acids: The Genetic Information Media and a Template for Nanotechnological Applications

While most readers are familiar with the double helical structure of the DNA (deoxyribonucleic acid) as was suggested by Watson and Crick in 1953 (for which they won the Nobel Prize in 1962), for quite a long time it was not considered as typical nanomaterial. Yet, DNA is a typical well-ordered structure in the nano-scale. It has a diameter of about 2 nm and a length that ranged from a few nm of many microns (the length of each base pair is 0.34 nm). The DNA is a self-assembled structure that forms spontaneously by two complementary stands in which cytosine (C) moieties interacts with guanine (G) moieties and thymine (T) interact with adenine (A). All of these structures occur through a process of self-assembly. The overall structure is being stabilized by aromatic base-stacking as appears in other bionanostructures as will be discussed later.

The ribonucleotide acid (RNA) can also form complex structures at the nano-scale. Yet, these structures are usually singled stranded structures in which intramolecular, rather than intermolecular interactions, direct the formation of complex three dimensional structures of RNA such as the transfer RNA (tRNA) molecule and structural building blocks of the ribosome nanomachine.

Much of the pioneering work on the use of DNA as a building material for nanostructures was done by Nadrian Seeman from New York University (Seeman, 2005). As will be described in more detail later in this book (Chapter 6), Seeman was able to the demonstrate that complementary DNA strand could be designed to form complex two dimensional and even three dimensional cubic structure at the nano-scale. DNA appears to be a very attractive material for nanotechnology due to its relative thermal stability, as compared to protein building blocks. Moreover, the genetic engineering revolution also advanced very significantly the field of DNA chemistry. It is possible now to synthesize very large DNA oligomers in high quantities, a variety of chemical modifications and a relatively low cost.

Another very important work that used DNA as a nanotechnological material used the DNA as a template for the fabrication of metallic nanowires (Braun *et al.*, 1998). This will also be further elaborated and illustrated in Chapter 6. Briefly, DNA molecules were coated with silver to form conductive nanowires. The same group went one step further to perform protein-based lithography on DNA templates (Keren *et al.*, 2002). In this case a DNA binding protein, RecA, was used to protect specific parts of the molecule from metal deposition and thus allow a sequence-specific modification of the DNA template. In principle by the design of exact recognition sites with a given DNA oligomer a very specific pattern of metallization could be obtained.

Small oligonucleotides can serve a structural and even functional role. It was demonstrated that small RNA molecules, known as ribozymes, can have enzymatic activity as in the case of some RNA elements of the ribosome. Later studies have confirmed the ability of other small RNA and DNA molecules to carry various enzymatic activities and the ability of small nucleic acid oligomers, know as aptamers, to serve as specific binders of various ligands.

3.7. Oligosaccharides and Polysaccharides: Another Class of Biological Polymers

The oligosaccharides and polysaccharides are polymeric chains in each sugar molecules which are joined together by the formation of glyocsidic bond upon the reaction of aldehyde or ketone of one sugar molecule with the hydroxyl group in another sugar. Small poly-sugar molecules are known as oligonsaccharides in either branched or non-branched fashion. For example, glycogen is a long oligomer that is composed of glucose monomers that are joined together in large and branched polymers. Most homogenic saccharides are used as source for energy conservation. Yet, complex polysaccharides, such as peptidoglycan in which saccharides are complex with peptide template, serve as mechanical structures with a nano-scale order that keeps the mechanical properties of bacterial cells. Different oligosaccharides also have roles in molecular recognition of the immune system and various bacteria, as specific antibodies are being raised against specific polysaccharides structures. The exploration of oligosaccharides for specific nanotechnological uses is still to be made, but it may actually represent a very important direction in the future.

3.8. Amyloid Fibrils as Self-Assembled Nano-Scale Bio-Assemblies

The self-assembled structures as described in this chapter are mostly functional assemblies in living organisms. Another class of self-assembled structures is actually a pathological one. Yet, it plays a central medical role as well as an emerging role in the field of nanobiotechnology.

Amyloid assemblies are nano-scale fibrils that are characteristic of a large group of diseases of unrelated origin. A partial list of amyloid-associated diseases includes Alzheimer's disease, Type II diabetes, primary and secondary amyloidosis, and Prion diseases (such as BSE – Bovine Spongiform Encephalopathy, also known as the "the mad-cow disease"). These assemblies could be found in various tissues and organs depending on the specific disease (e.g., in the brains of Alzheimer's disease patients or in the pancreas of Type II diabetes patients).

Ultrastructural analysis of the amyloid assemblies using electron microscopy (EM) and atomic force microscopy (AFM) indicate that the deposits have typical 7-10 nm diameter fibrillar morphology and a length of several micrometers. Furthermore, the observed fibrils are highly ordered with a clear meridian X-ray fiber diffraction pattern of about 4.6 - 4.8 Å. These canonical elongated structures have a diameter of about 7-10 nanometers, a β-sheet secondary structure, and a strong birefringence upon staining with the Congo Red dye. The regularity of the formed fibrils along their long axis, together with the classical nucleation and growth mechanism, has led to the reference of the process of amyloid formation as "one dimensional crystallization" (Jarrett and Lansbury, 1992).

While amyloid-fibrils nano-assemblies were previously considered to have only a pathological role, recent studies clearly demonstrated that the amyloid phenomenon may actually represent a significant physiological process. The first indication for such a role was the demonstration that the formation of biofilm by the *Escherichia coli* bacteria is facilitated by the self-assembly of the major curlin protein, CsgA, into typical amyloid fibrils (Chapman *et al.*, 2002). This finding was a milestone in the establishment of the notion that amyloid fibrils may act as functional physiological entities, and not only as pathological ones. More recent studies revealed a role of amyloid structures in the formation of aerial hyphae by the *Streptomyces coelicolor* bacteria (Elliot *et al.*, 2003). Amyloid self-assembly was also shown to have a role in the formation yeast prion (King *et al.*, 1997), a physiological phenomenon that may have a clear evolutionary role in yeast physiology (True *et al.*, 2004).

Amyloid fibrils were already demonstrated to serve as building blocks for nano-electronics as part of yeast prion protein which was genetically engineered and used as a template for the production of conductive nanowires. Engineered cysteine residues were used to bind gold nanoparticles along the self-assembled fibrils. Enhancement protocol was then used in order to get continues conductive nanowires (Scheibel *et al.*, 2003) as will be further illustrated in Chapter 6. The recognition principles as implicated in the formation of amyloid fibrils and most notably the key use of aromatic interactions were later used in the design and fabrication of nanotubes and nanospheres.

3.9. Silk: Natural Fibrillar Supramolecular Protein Assembly

The natural fibrillar supramolecular structures that are mostly related to amyloid fibrils, as discussed above, are those of spider and worm silk. These fibrils are predominantly composed of two protein molecules that create macroscopic filaments by precise molecular recognition and self-assembly process (Kubic, 2002; Jin and Kaplan, 2003). The notable properties of silk fibers are their outstanding strength and high flexibility. In spite of its non-covalent supramolecular nature, spider silk is considered to be on the one hand remarkably stronger than a steel filament of the same diameter but on the other hand highly flexible. These unusual mechanical properties are the basis for the stability and functionality of the spider-net and cocoon. This serves as a vivid example of the unique properties of supramolecular biomaterial, and its potential superiority over traditional organic and inorganic materials.

3.10. Ribosome: The Protein Assembly Line Instrument

The earlier examples in this chapter are mainly concerned with the formation of assemblies that either have a structural role or carry genetic information. Yet, the process of biological self-assembly can actually form functional machines. The most complex bio-nanomachine whose structure is known at the atomic resolution is the ribosome, the protein-RNA hybrid machine for the production of new proteins. The ribosome uses encoded data in the form of messenger RNA (mRNA) to form specific proteins, a process known as translation. The information for the identity of the protein is encoded in a four letter - A, C, G, and T - triplet codes (known as codons). The 64 different combinations of the four letters represent degenerative multiple code table coding for the different 20 amino-acids plus three codons that signal a stop of the translation process. The coding molecule also includes information about the region of synthesis initiation.

The ribosome is a very complex self-assembled complex. The ribosome is complex structure with a diameter of about 200 nm. A ribosome consists of two subunits that fit together. Each subunit consists of one or two very large RNA molecules (known as ribosomal RNA or rRNA) A typical bacterial ribosome is composed of 55 different proteins and three RNA molecules. This complex functional assembly is formed by a spontaneous process of self-association. Indeed, if purified components of the ribosome are mixed together in a test tube, a functional ribosome can be formed. The route for self-assembly is a hierarchical and coordinated one. First specific proteins bind the very large RNA molecules, which is followed by interactions of other proteins to finally form a ribosome of the original structure as well as function. This process resembles in some ways the self-assembly of the much simpler TMV virus as discussed in Chapter 3. As in the case of TMV assembly some molecules, namely RNA molecules in both cases, serve as a template for further coordinated but spontaneously energy-favorable assembly of much more complex structure.

3.11. Other Complex Machines in the Genetic Code Expression

As mentioned above much is known about the main protein synthesis machine. Yet the order processes of transfer and duplication of genetic information are also mediated by rather complex assemblies. Various molecular machines mediate the duplication of the biological information media, the DNA, as well as its transcription into the ribosome input the mRNA. Cellular proteins, known as transcription factors, serve as on and off switches for the expression of this data according to the correct external conditions or specific differentiation of the cell in the case of complex multicellular organisms. Other cellular machines are responsible for the maintenance of the correct DNA composition by error-correction mechanism. These complexes are responsible, for example, for the correction of radiation induced damage of the DNA. Indeed, individuals whose correction machines are dysfunctional suffer from a high incidence of cancer.

3.12. Protein Quality-Control Machinery: The Proteosome

It is important to note that the formation of protein structures is not a one way process. The proteins that are synthesized by the ribosome in the process of genetic information expression are dynamic biochemical species. Protein can be degraded by the cellular machinery into its amino-acid building blocks. The half life of a typical protein ranges from few minutes to several days. The degradation of the proteins is a well-coordinated process that is mediated by complex cellular machinery as was discovered by Hersko, Ciechanover, and Rose.

Proteins that are destined for degradation are marked by a molecular protein tag known as ubiqutin. The tagged, depicted as ubiqunated, proteins are then degraded by a complex molecular assembly known as the proteosome. The proteosome is a nano-machine that has a shape of an elongated cylinder with a diameter of about 10 nm and a length of several tens of nanometers. The proteosome accepts ubiquitineted protein substrates and use ATP energy to break them into basic building blocks.

3.13. Biological Nano-Motors: Kinesin and Dynein

Biological systems can form nano-motors that are ordered of magnitudes smaller than any man-made ones. Two biological motors that were mentioned in Chapter 3 are Kinesin and Dynein, known as Microtubule Associated Proteins (MAPs). These two biological motors travel along the microtubules to transport organelles, vesicles, or other components of the cytoskeleton along the tubules (Marx *et al.*, 2005). These transported entities are known as "cargo". The molecular motors move on the tubules like trains over railway tracks. As described later, each of the motors is moving in different directions along the microtubules. Kinesins move cargo away from the centrosome (anterograde transport), while dyneins transport cargo toward the centrosome (retrograde transport).

The energy needed for the movement of the motor proteins along the micro tubules track in supplied by ATP molecules. The kinesin motor is much smaller and less complex than dynein, with a molecular weight of 380,000 and over 1,000,000 for kinesin and dynein, respectively. Therefore, the kinesin complex was studied extensively. The kinesin molecule is organized into three domains: two large globular head domains which bind ATP and facilitate the movement, that are connected by a long coiled-coil helix to a tail domain, which binds the cargo. Experiments that were done with purified kinesin and microtubules had revealed the molecular mechanism of movement. It was found that the motor moves in discrete 8 nm steps. Mechanical measurement had revealed that the motor exert a force of about 6 pico-Newtons (pN, 10^{-12} Newtons).

3.14. Other Nano-Motors: Flagella and Cilia

Another form of mechanical movement in biological systems is the one related to the movement of cilia or flagella. These types of swimming devices are common to bacteria, protozoa, and sperm cells. The cilia-mediate movement of eukaryotic organisms is based on the same molecular machinery as the microtubules motor with dyneins as the major motor unit. Also in this case chemical energy is being converted to mechanical energy at the nano-scale to support the movement of various elements.

Another type of biological nano-motor is the F0-ATPase motor. This type of motor is found in bacteria as well as in mitochondria, the energy powerhouse of eukarytic cells, which is of prokaryotic origin. This protein complex is used for the synthesis of ATP in the case of mitochondria. A proton motive force direct a forward movement of the motor that results in the production of ATP. In the case of bacteria, the very same motor structure allows the production of mechanical movement for which the energy is supplied by ATP hydrolysis.

3.15. Ion Channels: Nano-Pores of High Specificity

Another group of nano-scale functional assemblies are the membrane embedded ion channels that were already mentioned earlier in this chapter. This class of protein assemblies forms, by a process of self-assembly within the membrane, a nano-scale filter that allows very specific transfer of a specific type of ions but not others. The most common ion channels are the sodium channels, potassium channels, and calcium channels. These channels have a key role in the electrophysiological process in which the function of the muscle is being controlled by the central nervous system.

Much of our current knowledge about the structural basis of the remarkable selectivity of ion-channels at the nano-scale comes from the pioneering work of Roderick MacKinnon. MacKinnon, who was the first to crystallize and determine the high-resolution structure of an ion channel (Morais-Cabral *et al.*, 1998) was awarded the 2003 Nobel Prize in Medicine for his outstanding achievement. This was followed by the structural determination of several more ion channels assemblies. The structural information together with biochemical and biophysical data on the function of the channels should allow the design of these selective nano-scale filters in the future.

These molecular assemblies are ubiquitous in nature and could be found in all organisms. In higher organisms there are multiple forms of each type of channel that could be modulated by various signals, including physical (voltage gated) and chemical (ligand gated). Thus these are assemblies which can indeed be considered as a nano-scale devices that open and close very precisely at the nano-scale upon external control. Due to these properties of the channels, there are many nanotechnological attempts to either use existing channels for various detection and sensors in the nano-scales. Other attempts are directed towards the design of artificial nano-scale channels.

The formation of pores within the membrane is also being used by various organisms as a toxic strategy (Parker and Feil, 2005). Such pore forming toxins permeate the membrane by the formation of nano-scale pores that are embedded within the lipid membranes. The aberrant flow of ions across the pore finally results in cellular dysfunction and often cell death. Such toxin-based nanoporse were already used for the nanotechnological application (Meller *et al.*, 2005). For example, the 2 nm α-hemolysin pore was used in the transfer of nucleic acid polymers across membranes. The application of transmembrane potential serves as a driving force for the translocation of the charge DNA or RNA molecules. The monitoring of the changes in the current as is reflected during the transfer of the oligonucleotides across the pore is envisioned for rapid sequencing of DNA sample that should have many applications in genomics and diagnosis (Meller *et al.*, 2000; Mathe *et al.*, 2005).

Chapter 4

Nanometric Biological Assemblies: Molecular and Chemical Basis for Interaction

The specific interaction between building blocks, a principle that was already put forward in this book, is the key for the execution of the assembly processes. In this chapter, a brief introduction to the chemical and molecular basis of recognition and association processes is presented. We would like to give some formal background regarding the molecular and chemical basis for the introduction of these concepts to readers with no formal background in chemistry. We will discuss the thermodynamic basis for interactions and consider its entropic and enthalpic parameters. We will also provide some insights on the kinetics of macromolecular complex formation and its relevance to the biological function.

4.1. Emergence of Biological Activity Through Self-Assembly

The process of self-organization was already discussed in previous chapters. In the context of biological nano-machines and functional assemblies, it is related to the way in which each key complexes and elaborate functionality are being produced by spontaneous association of discrete biomolecular elements into well-ordered assemblies. The association of biological assemblies is usually sequential and not parallel. Thus high affinity, the equally important remarkable specificity, between biomolecular motifs is essential for the formation of these structures by sequential processes of molecular recognition and self-assembly that leads to the formation of complex nano-scale machines.

The recognition motifs are usually protein molecules or nucleic acids (both DNA and RNA), although as discussed earlier in this book biological affinity and specificity could be facilitated by other biomolecules such as sugars and its polymeric form of polysaccharides. We will first discuss the basic concepts of molecular recognition, affinity, and specificity from the chemical point of view. This will serve as the theoretical basis for the understanding and the experimental determination of the processes of recognition and assembly that lead to the formation of the ordered structures as described here. In the next chapter, we will put these chemical rules in the context of biological recognition and the formation of specific biological assemblies such as nucleic acid complexes and receptor-ligand interaction.

4.2. Molecular Recognition and Chemical Affinity

The process of molecular recognition is a key concept in the study of supramolecular chemistry as the chemical basis for non-covalent interactions between defined molecular moieties. This is of course not limited to biological systems but describes the interaction between chemical entities in general. The term "molecular recognition" describes the ability of two or more molecules to interact with each other at high affinity and specificity. As was described in Chapter 2, it is one of the cornerstones of supramolecular chemistry which is by itself one of the major aspects of nanotechnology.

The affinity between discrete molecular assemblies is often defined by the dissociation constant (K_d) of the formed complex. The value of K_d quantitatively describes the intensity of the binding between two or more molecules. This value, that is constant for a given molecular system, is a thermodynamic equilibrium value that could be described by the following formula for a simple two-component systems in which two molecules A and B form a dimeric complex (obviously in the case of multiple component system, the mathematical expressions become more complex):

$$K_d = [A] \times [B] / [AB] . \tag{4.1}$$

Where [A] and [B] are the molar concentrations in equilibrium of molecules A and B, respectively. While [AB] is the molar concentration of a bimolecular complex composed of A and B. As could easily be observed from the equation, the units of K_d for this simple bimolecular system are molars (abbreviated as M). The K_d is often attributed as to the "affinity" of the macromolecular system. Thus an "nM affinity" means a K_d value of nanomolar (10^{-9} M).

It is worth noting that for many practical applications, it is often complicated to determine the exact concentration of all free molecular species. However, if one of the molecules, lets assume B, is much lower than the concentration of A (i.e., [B]<<[A]), one can assume that at a given concentration the amount of added A ($[A_{total}]$) is almost equal to the amount of free A (denoted as [A]). Under these conditions the concentration of total A in which half of B is bound equals to the K_d, as under these conditions:

$$[B] = [AB] \rightarrow K_d = [A] \sim [A_{total}]$$

This particle approach is often used to experimentally determine the values of K_d in a relatively simple way. Allowing a large difference between the concentration of A and B, molecule B is being labeled with very sensitive probe such as radioactive or high quantum-yield fluorescence labeling. Such experiments were used to determine the affinity between antibodies and antigens, DNA and DNA-binding protein, and receptors and ligands as will be described in the next chapter.

Specific binding is usually attributed to K_d values that are lower 10^{-6} M (1 μM), although as will be described below for specificity the relative affinities are more important than the nominal value. High specificity binding as between antibody and antigens or transcription factor and DNA are often in the nanomolar (10^{-9} M) range. However the affinity could be even higher, as some antibodies-antigen complexes have K_d values as low as 10^{-11} M. The strongest non-covalent biomolecular complex is that between avidin and biotin that is frequently used as a connecting adaptor in biotechnological and nanotechnological applications (Wilchek and Bayer, 1990). Since this avidin-biotin binding is so strong, it could not be determined accurately, but it is assumed to be in the order of 10^{-15} M which practically behaves as a covalent binding between atoms.

4.3. Affinity and Specificity of Biological Interactions

A strong affinity between two biomoelcular entities is of course important but yet not sufficient for high specificity. The specificity of biological interaction is defined by the relative affinity of a given molecule to its molecular target as compared to non-specific targets. For example in the case of DNA-binding proteins, that are often positively charged molecules, there is some non-specific interaction even with non-specific DNA fragments that are negatively charged polyanions. Yet, while the non-specific DNA-protein interactions are in the order of 10^{-3} to 10^{-4} M, the specific binding of the proteins to the correct molecular target is often in the range of 10^{-7} to 10^{-9} M. There are also other cases in which the same molecular receptor can bind different ligands in various affinities and thus modulate the biological activity of the complex by some combinatorial binding pattern.

A clear specificity is not limited only to the binding of simple complexes but can also be related to the specificity of ion channels, the transmembrane nano-channels that were described in Chapter 3, that allow the transport of specific ions but not others. For example, the potassium channel that allows the effective transport of potassium ions but not the very similar sodium ions. The specificity is reflected by orders of magnitude difference in the transport rate as compared between the two cations. Another biological specificity stems from the activity of enzymes that can distinguish between very similar substrates and catalyze unique chemical reactions. An example for this specificity is the activity of sugar modifying enzymes, that are critical for energy conversion cascades such as the cellular glycolysis, which specifically and clearly distinguish between similar sugars such as glucose and fructose.

4.4. The Relation Between Thermodynamics and Kinetics of Dissociation

While K_d is a thermodynamic parameter, it is of course directly linked to the kinetic properties of the complexation reaction (for a simple bimolecular dimeric system):

$$K_d = k_2/k_1 \qquad (4.2)$$

Where:
k_1 (rate of dissociation) **AB ---> A+B, and**
k_2 (rate of association) **A+B --->AB**

Thus, as k_2, the rate of association, is related to the concentration of the components of the complex, k_1 is independent of the concentration of the complex. While the rate of association is usually similar for various biomolecular partners of similar size and is directed by the rate of diffusion, the rate of dissociation is directly linked to the affinity of the complex. Due to this property, in many cases in which there is a problem to have one of the molecular partners in a much lower concentration as compared to the other one (as described in Equation 4.2), the rate of dissociation serves as a criterion for the molecular affinity. Many high affinity complexes may have dissociation rates of hours or days, while low affinity complexes may have dissociation rates of milliseconds or shorter.

If specific antibodies are available, the kinetic rates could be determined using common biological assays such as ELISA (enzyme-linked immunosorbent assay) in which the ligands are attached to multi-well plastic plates and the interaction is probed by antibodies. In other cases, the degree of complexation could be determined by colorimentric methods or fluorescent spectroscopy. However, in recent years there has been an increased tendency to use the surface plasmon resonance (SPR) technique for the determination of the binding kinetics between non-labeled species. There are several commercially-available instruments that utilize the SPR technique, the most famous one is the Biacore®, but other commercial systems from other manufactures are available.

The basic principle of this technique is based on the mobilization of one of the interacting molecules into the surface and measurement in the changes in the optical properties of the surface upon the binding of the second molecule. The surface plasmon is an electromagnetic surface-bound wave that propagates along the interface between a metal and a coating layer. In the commercial instruments, a metal surface is modified with biological molecules and the changes in the reflective index upon binding are determined (Figure 4.1).

During a typical SRP experiment, a solution that contains the binding molecules is injected to the chamber that contains the other partner of the molecular pair that is bound to the surface. The real-time change in the resonance signal as a function of time is plotted. This plot that is referred to as a "senosgram" gives information about the rate of association of the molecules (Figure 4.1). When the injection of the molecules is halted and the only vehicle solution is being injected, a dissociation process occurs, which could also be followed. As discussed above, when both the association rate constant as well as the dissociation rate constants are known, the thermodynamic K_d could be directly calculated. While the SPR is a very useful method, it should be noted that it is useful when the immobilization of the molecules to the surface do not affect the binding reaction and when the injected molecules does not interact with the surface by non-specific adhesion or other interactions.

4.5. The Chemical Basis for Molecular Recognition and Specific Binding

The formation of complex non-covalent biological structures often stems from the combination of many small energy interactions that the linear combination of their contributions together forms a complex structure of high affinity. Such interactions can include hydrogen bonds (relative energy of less than 1.0 Kcal/mol), var der walls interactions, hydrophobic interactions, and electrostatic interactions. The overall free energy of interaction can be derived from the dissociation constant as:

$$\Delta G = RT \ln(K_d) \qquad (4.3)$$

Where ∆G is the free energy of interaction, R is the gas constant, and T is the absolute temperature.

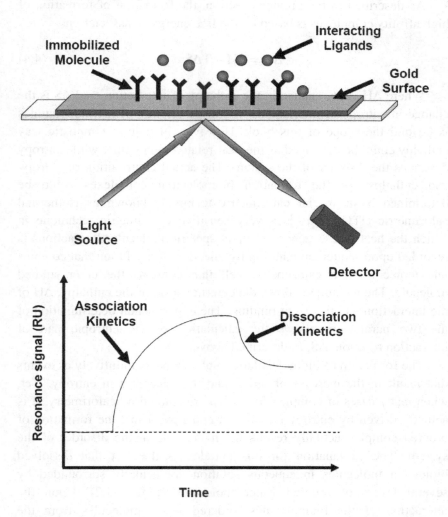

Figure 4.1: The use of SPR to monitor the kinetics of association and dissociation between biomolecules. The interaction between the molecules on the metal surface affects its optical properties. When the signal is plotted as a function of time, the kinetics of association and dissociation could be resolved.

4.6. The Formation Specific Complexes by an Increase in Entropy

As described in the previous section, the formation of formation of high affinity complexes is based on the free energy of association.

$$\Delta G = \Delta H - T\Delta S \qquad (4.4)$$

Where ΔH is the change in the enthalpy of interaction, and ΔS is the change in entropy of interactions. A full description of thermodynamics is beyond the scope of this book. However, in quite a simplistic way enthalpy could be described as the heat-related interaction, while entropy describes the disorder of the system. The actual contribution of entropy and enthalpy for the formation biomolecular complexes could be determined by a special calorimetric technique known as isothermal calorimetric (ITC). This is a very sensitive calorimetric technique in which the heat that is generated upon specific molecular interactions is recorded upon sequential addition by injection of a solution that contains one molecule to a calorimetric cell that contains the other studied molecule. The technique allows the determination of the enthalpy, ΔH of the interaction and the K_d of binding. The experimental determination of the two parameters allows the calculation of the entropic part of interaction reaction, ΔS, as described above.

The formation of biomolecular complex appears intuitively as events that result in the increase of order, and thus decrease in entropy. Yet, when many cases of complex formation are studied by calorimetry, it is actually driven by entropy rather than enthalpy. Thus, the formation of ordered complex actually results in an increase in the disorder of the system. The explanation for this paradox is the fact that dissolved biological molecules in aqueous solution are actually surrounded by several layers of ordered water molecules (Figure 4.2). Upon the interaction of the biomolecules, ordered water molecules from the recognition interface are released to the bulk water which results in an overall increase in the disorder of the system.

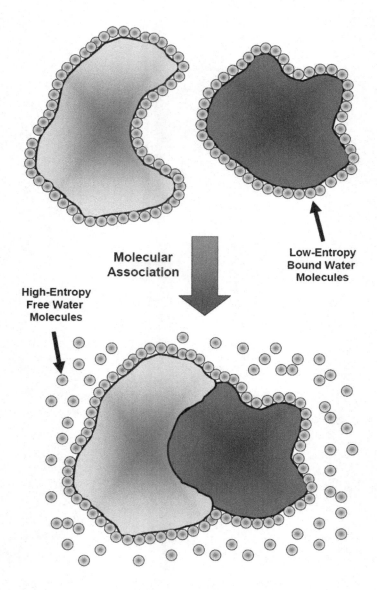

Figure 4.2: The molecular basis for association between macromolecules that is driven by entropy.

Chapter 5

Molecular Recognition and the Formation of Biological Structures

In the previous chapter, we had provided some theoretical background on the kinetic and thermodynamic basis for the specific recognition between macromolecules. We had discussed some formal aspects of the association and dissociation both as a dynamic kinetic process as well as an equilibrium thermodynamic state. In the current chapter we will describe a few of the specific recognition and assembly processes that lead to the formation of biological complexes. We will describe both the mechanism of formation of the biological complexes as well as some possible directions for their use in nanotechnological applications.

5.1. Antibodies as the Molecular Sensors of Recognition

The ultimate cellular entities which allow specific recognition of various compounds, organic and inorganic, biological and non-biological, are the antibody proteins. The unique recognition properties of these relatively large protein molecules allow very high affinity of molecular recognition of various ligands. The K_d of the interactions of antibodies with a given ligand can reach the order of 10^{-12} M (picomolar) or even femtomolar (10^{-15} M) concentrations. Antibodies do not only possess high affinity, but also high specificity as antibodies can distinguish between extremely similar ligands. Antibodies are of course a key component of the immune system and have a crucial role in the protection of the body against infections and also malignant transformation. The description of the mechanism of the immune system is beyond the scope of this book, but those readers with no background in immunology are highly encouraged to read further about this remarkable system.

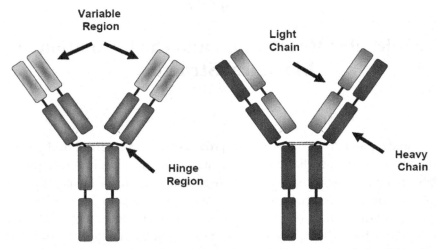

Figure 5.1: The basic structure of antibodies. The Y-shape structure is formed by the interaction of a heavy chain and a light chain. The specific recognition of the antibodies molecules is facilitated by the unique structural features of the variable region.

Just as a brief overview, we would like to mention that the high affinity of antibodies is achieved by selection of specific recognition interfaces for a set of complex combinatorial amino-acid arrangements (Figure 5.1). Typical Y-shape structure of an IgG (Immunoglobulin G) antibody is made up of V (for variable) and C (for constant) regions (Figure 5.1). A hinge region in the middle of the IgG antibody confers flexibility on the molecule. The high binding of antibodies to their ligand is further increased by the avidity of the antibodies. The cellular antibodies contain more than one binding site for a given ligand. In the case of the IgG antibodies the affinity is increased by the existence of the two binding sites. There could be up to five identical recognition sites for the IgM antibodies. In this way, much stronger binding is achieved. The effect of the multiple binding sites could be intuitively understood by the consideration of the dissociation rate constants that were described above. In order for the antibody to be released from its molecular target, there is a need for release from the two binding sites at the same time. As the two or more dissociation steps are statistically independent events the probability for antibody release is the product of the independent probabilities.

For the purpose of this book, we will describe the technological application of antibodies for various applications. The biotechnological uses of antibodies were already mentioned in Chapter 1, as some of the common diagnostic kits such as home pregnancy-test kits are based on the use of the high specificity of antibodies and their ability to detect minute amounts for ligands in a complex mixture. The combination of antibodies together with nanotechnology tools could lead even to higher levels of sensitivity. It was demonstrated that the conjugation of specific antibodies with DNA bar-code magnetic nanoparticles could lead to atto-molar (10^{-18} M) scale diction of the prostate-specific antigen (PSA) protein, a key indicator of prostate cancer (Nam *et al.*, 2003). This is a clear example of the great prospects of the combination of biotechnology together with nanotechnology for the ultra-sensitive bio-sensing. In the case of PSA this could lead to early detection of prostate cancer, yet the attainment of ultra-sensitivity could have ample application outside of the medical field. For example, in environment monitoring and homeland security applications in which minute amounts of biological and chemical agents could be detected.

5.2. Selection of Antibodies and Equivalent Systems in the Test Tube

As mentioned above, antibodies are one of the most elaborate biological molecular recognition modules. The excellent affinity and specificity properties of the antibodies encouraged the attempts to develop such systems in the test tube. Artificial libraries that contain antibody repertoires are being expressed by bacteria and after cycles of selections, high-affinity antibodies are being selected. This allows us to obtain antibodies that have affinity and specificity and also to "self" ligands or other entities with low antigenic potential. The field of antibodies selection has advanced significantly in recent years. One of the most important directions is the engineering of single-chain antibodies. These antibodies should have many uses in further therapeutic applications.

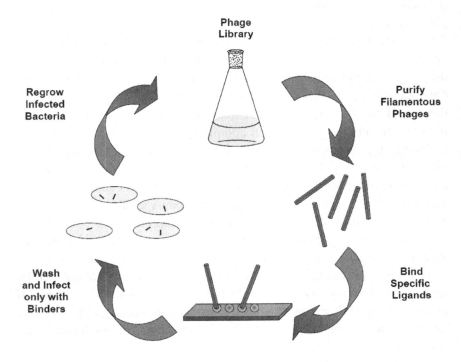

Figure 5.2: Selection of specific binders by phage display techniques. Iterative cycles of enrichment are being used to get high-affinity binders. The selection starts with a random library which is significantly enriched after only binder phages are retained and further used to infect producing bacteria.

Another mean of achieving high-affinity biological systems is the use of phage display systems (Figure 5.2). In those systems, antibodies or other peptidic entities are presented on the surface of filamentous bacteriophages. Several cycles of binding and selection, known as panning, allow us to get phage particles that bind the molecular target at high affinity and specificity. Another direction for obtaining molecular recognition entities from non-living systems is the use of chemical peptide libraries. These libraries could have molecular diversity of 10^{-12}

or even higher that can allow us to obtain small molecules that bind their target at relatively high-affinity. While the disadvantage of the peptide libraries is the inability for sequential enrichment using cycles of panning, one of the advantages is the ability to use non-natural building blocks in the libraries. This of course allows us to significantly increase the chemical diversity and to obtain non-cleavable peptide entities. This is especially important for the therapeutic industry. The use of synthetic chemical libraries is of course not limited to peptides. Other organic, or even inorganic molecules, could be probed for their ability to bind various bimolecules.

The use of bacteriophages as building blocks for nanotechnology was already demonstrated by the pioneering work of Angela M. Belcher and her co-workers. This will be described in detail in Chapter 6. Briefly, the nano-scale size of the bacteriophage building blocks together with the ability to engineer them with specific recognition abilities allows the formation of nanostructures with unique recognition properties.

5.3. Recognition Between Nucleic Acids by Proteins

The recognition of specific DNA sequences by protein molecules known as DNA-binding proteins was already mentioned earlier in this book. This specific recognition lays the basis of the complex control of the expression of the genetic information in various organs and tissues in multicellualr organisms and under different growth conditions by microorganisms.

The DNA binding proteins have the unique ability to recognize a specific sequence motif that could be as short a 6 or 8 basepairs out of the millions or billions of basepairs in a given genome. In order to execute a correct molecular response, these proteins have not only to bind with high affinity regulatory DNA motifs but the specificity of interaction needs to be high. The complex cascade of protein-DNA interaction that controls a biological complex was first demonstrated in the lactose operon of *E. coli*, but was later described in much more complex organisms including human cells.

One of the fascinating properties of DNA binding proteins is their ability to recognize their specific binding sites very quickly and accurately out of millions of possible communications. Although several high resolution structures of protein-DNA complexes are already known, the exact way in which the protein finds its specific target is not fully understood. It is assumed that some non-specific binding of the protein to the DNA occurs by non-specific electrostatic interactions by either scanning the DNA strand or randomly moving from one site to another until a high affinity binding, which is reflected by very slow dissociation, occurs. Interestingly, careful biophysical analysis had indicated that the specific protein-DNA interaction may be facilitated by entropy rather than enthalpy (Garner and Rau, 1995). Thus, the non-specific initial binding is mediated by electrostatic interactions between the positively charged DNA-binding proteins and the polyanion DNA, while the specificity is modulated by the release of hundreds of water molecules upon the specific binding into the exact recognition site. The molecular basis for such mechanisms for specificity is explained in Chapter 4.

DNA binding proteins such as the RecA protein were already used for nanotechnological application as will be described in Chapter 7 Briefly, the ability of the protein to bind a very specific site in the DNA with sub-nm precision, is used for bio-lithography where the protein serve as a nano-scale mask on the DNA.

5.4. Interaction Between Receptors and Ligands

Molecular recognition also serves the execution of biological function in the extracellular milieu. Various receptors are embedded in the cellular membrane and allow the response of the cells to various signals. The best studied receptors are those that recognize hormone signals. Hormones are used to carry biological signals through the blood stream and the recognition by the membrane receptor allows the correct biological response.

A paradigm for hormone receptors is that of insulin receptor. The level of blood insulin is probed by membrane imbedded receptors in the periphery tissues. The recognition is mediated by a set of non-covalent interactions as described above for other supramolecular assemblies. The high affinity between insulin and its receptor allow sensitive sensing but not too long occupancy of the receptor site. The binding of the insulin to the insulin receptor results in the executing of a molecular cascade that ends up in the activation of specific genes that regulate the metabolism of sugars in the target cells.

The interaction between receptors and ligands may also have nanotechnological applications as the high affinity and specificity can allow the accurate assembly of subunits into nano-scale order assemblies, the modulation of activities of nano-devices and assemblies, and the use of these interactions for diagnostic purposes.

5.5. Molecular Recognition Between Nucleic Acids

One of the most important specific recognitions between biological molecules is the interaction of complementary DNA and RNA strands. As most of the readers know, DNA is formed by four types of bases A, C, T, and G, and the molecular basis for the high specificity of DNA double strand structures derives from the specific affinity between A and T pairs and those made of C and G. The energy of interaction between single base pairs is quite small (three hydrogen bonds stabilize the CG pair and two hydrogen bonds stabilized the AT pair). Yet in a typical double stranded DNA when tens, hundreds, or thousands of bases pairs are joined together, the overall chemical affinity is quite high.

DNA can also form specific hetero-molecular complexes with RNA molecules. These interactions are based on the same recognition rules as described above. Yet, in the case of RNA molecules, the T base is substituted by the U base to form specific AU basepairs. RNA, and in some cases DNA, can form intermolecular basepair interactions to form complex three-dimensional structures in the nano-scale. The best known example is the formation of the transfer RNA (tRNA) molecules.

The specific interaction between nucleic acid molecules is already used in several bionanotechnological applications. The specificity of nucleic acid interaction can serve as a very specific addressing entity or molecular barcode. Moreover, DNA molecules by themselves can serve as a template for the formation of two dimensional and three dimensional assemblies as was demonstrated by the pioneering work of Nadrian Seeman that will be described in Chapter 6.

Self-Assembly of Biological and Bio-Inspired Nano-Materials

In the previous chapters, we described the formation of nanostructures in various biological systems, the thermodynamic and kinetic basis for the molecular recognition and self-assembly process, and some key examples for such recognition and assembly processes in physiological-relevant systems. In the current chapter we will describe the design of novel biological and bio-inspired building blocks that are based on the principles that were described earlier. These designed nanostructures have a great potential for both bionanotechnological and nanobiotechnological applications.

6.1. Formation of DNA-Based Materials

One of the earliest studied assemblies of ordered materials at the nano-scale was that of two dimensional and three dimensional DNA structures. The physiological basis for the interaction of DNA at the nano-scale was described in Chapter 2. Here we will describe the use of these principles. The pioneering work in this field was made by Nadrian Seeman already few decades ago before nanotechnology became such a popular field (Seeman, 1990, 2005). Seeman, a crystallographer by training, had recognized the great potential of the molecular recognition and complementary of DNA oligomers, as were described in Chapter 4, to fabricated complex materials that are based entirety on molecular recognition and self-assembly processes (Figure 6.1). Long before nanotechnology became such a "hot topic" as we know it today, Seeman already manipulated DNA assemblies at the nano-scale to form complex and elaborated structures (Seeman, 1990).

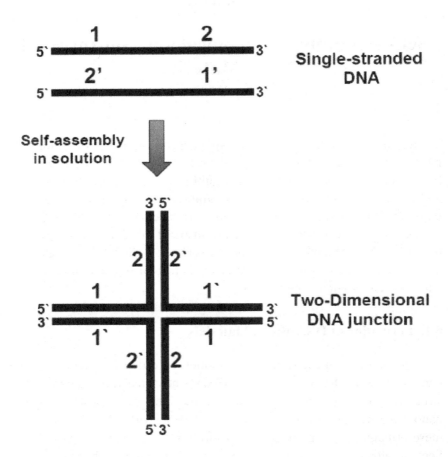

Figure 6.1: Self-assembly of linear DNA polymers into two-dimensional junctions. The interaction is based on the complement nature of the single-stranded DNA molecules (1 and 1', 2 and 2').

Seeman had used DNA oligomers that have internal complementary recognition interfaces (Figure 6.1). When those DNA molecules were mixed in aqueous solution, a spontaneous self-assembly process occurred that resulted in the formation of order two dimensional arrays of DNA. Follow up studies had made the design into even more complex two and three dimensional structures such as DNA cubes. A recent study from the group had also provided the DNA architecture to design DNA-based nanotubes (Sherman and Seeman, 2006).

Since the early days of DNA-based materials many applications by various groups were developed (for a recent review see Feldkamp and Niemeyer, 2006). DNA oligomers have many advantages as molecular materials due to their high thermal stability as compared to proteins (but not to peptides as will be discussed later in this chapter). Furthermore, the molecular biology revolution had created a great need for synthetic DNA oligomers for various reactions. This resulted in the optimization of the large-scale solid-phase chemical synthesis of DNA oligonucleotides (known in short as "oligos") and a significant decrease in the price of custom-made DNA. The price of DNA oligomers was reduced in orders of magnitude over a period of 20 years. Moreover, the need for various modified versions of DNA increased the chemical diversity of the commercially available DNA oligos. One of the disadvantages of the DNA assemblies is their sensitivity to enzymatic degradation by various nucleases that ubiquitously presented on most surfaces.

DNA is not only used by itself for molecular assembly of nanostructures but also as a template for the organization of other molecules such as proteins and peptides. Christof M. Niemeyer and coworkers as well as Wolfgang Pompe and coworkers had developed very intriguing applications of the specific recognition of DNA for the organization of proteins and inorganic nanoparticles to form various functional nanodevices (Simon *et al.*, 2004; Feldkamp and Niemeyer, 2006). One of the interesting applications involves biotinylated DNA oligonucletides that specifically bind avidin-modified surfaces. The remarkably high affinity between avidin and biotin results in extremely efficient organization of the molecular building blocks into highly ordered assemblies at the nano-scale. Such nano-patterned surfaces were also further modified by metal coating for achieving functional electronic devices.

6.2. Peptide-Based Nanomaterials

As the DNA molecules or oligomers, peptides represent another major group of biological macromolecules. Peptides which are polymers of amino-acids are some of the most important and diverse biomolecules. Synthetic peptides, which can be as short as dipeptide but also as long as 40-mers polymers, are most important and chemically diverse entities for nanotechnology applications (Zhang, 2003; Reches and Gazit, 2006). Like DNA oligomers, peptides are also synthesized readily and efficiently by solid phase methods. Actually the chemical synthesis of peptides was demonstrated well before the synthesis of DNA oligomers. Already in the 1960s, Bruce Merrifield and coworkers had developed the solid phase synthesis of peptides that has been practiced using similar principles until the present time (Merrifield, 1965). For this seminal work, Merrifield won the 1984 Nobel Prize for Chemistry.

One of the great advantages of peptides is their chemical diversity as compared to nucleic acids oligomers. DNA oligomers are basically composed of only four different letters. Therefore the chemical complexity of a given DNA oligo is 4^N, where N is the length of the oligonucleotide. On the other hand, complexity of peptides that are composed of only the 20 natural amino acids is 20^N. Therefore large diversity is achieved even with very short peptides. For example the chemical complexity of natural pentapeptides is $20^5 = 3,200,000$. The complexity of a 5-mer DNA oligomer is only $5^5 = 3,125$, more than three orders of magnitude lower. The actual situation is even more extreme as there are hundreds of non-natural amino acid monomers that are commercially available. Moreover, many termini modifications also exist. These modifications can significantly affect the biochemical and biophysical properties of the peptides. Therefore, the practical chemical space of a simple pentapeptide that contains non-natural amino acid is more than 10^{10}.

Peptides can also be synthesized in a kilogram scale in quite a simple way and inexpensively for short ones. The best example of a dipeptide that is synthesized at very large scale is the aspartyl-phenylalanine methyl ester, also known as Aspartame™. This widely used sweetener is synthesized at a cost of few cents per gram.

6.3. The First Peptide Nanotubes

Peptides were already recognized as excellent building blocks for nanotechnology already in the 1990s. Peptides have the great advantages of chemical diversity and flexibility as described above but also other advantageous properties for nanotechnology, including biocompatibility, and stability.

One of the key pioneering studies on the use of peptides as structural elements at the nano-scale was the synthesis of the first peptide nanotubes by M. Reza Ghadiri and coworkers (Ghadiri *et al.*, 1993). The researchers had demonstrated the formation of hollow nanotubular assemblies by the self-assembly of cyclic peptides, designed with an even number of alternating D- and L-conformation amino acids (Figure 6.2). The design was based on the unique architecture of the alternating isomers that is observed in some natural antibiotic peptide. The special design of the peptides resulted in flat ring-shaped subunits that are stacked together through intermolecular hydrogen bonds in a β-sheet conformation (Ghadiri *et al.*, 1993, 1994; Bong *et al.*, 2001). This conformation of stacked β-sheet structure is similar to that of nano-scale amyloid fibrils that were mentioned in Chapter 3.

The closed cycle and the alternating D- and L-conformations direct the side chains outwards from the ring and the backbone amides approximately perpendicular to the ring plane. The self-assembly process and the geometry of the nanotubes were shown to be controlled by rationally changing the number of amino acid in the ring, and the amino acid identity. The internal diameter of the nanotubes ranges between 0.7-0.8 nm and can be controlled by changing the number of the amino acids in the cyclic peptide sequence.

Follow up studies by the same group demonstrated the ability to engineer the side chains of the peptide nanostructures (Bong *et al.*, 2001). This is especially important for the design of conductive nanotubes by the correct orientation of organic systems that allow electron transfer in space through non-covalent systems (Horne *et al.*, 2005). These could lead to very important molecular nanoelectronics applications of these peptide nanostructures. Another practical application that was already demonstrated for this class of peptide nanotubes is their activity as antibacterial agents. This activity is achieved by the formation of nano-channels in the bacterial membrane by the nanotubes. This activity is similar to the effect of natural antimicrobial peptides on bacterial viability.

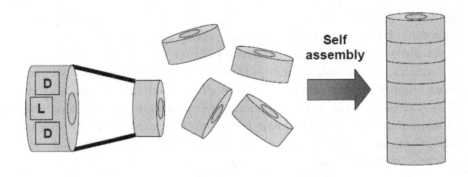

Figure 6.2: The formation of peptide nanotubes by the self-assembly of alternating L- and D-amino-acids. The peptide design resulted in the planar structure that allows the stacking of the peptides to form tubular hollow assemblies at the nano-scale.

Another cyclic peptide that was reported to self-assemble into well-ordered tubular structures is the Lanreotide octapeptide which is naturally synthesized as a growth hormone inhibitor. Ultrastructural analysis revealed that the Lanreotide peptide self-assemble into tubular structures with nanometric dimensions in aqueous solution (Valery *et al.*, 2003). These nanotubes are suggested to self-assemble through segregation of the aliphatic and aromatic residues that contribute to the conversion into β-sheet fibrils. The use of aromatic interactions with much smaller peptides, as simple as dipeptides, will be described further below.

6.4. Amphiphile and Surfactant-Like Peptide Building-Blocks

Another direction for the assembly of ordered peptide structures at the nano-scale is to use the same physical principles that drive the formation of phospholipids membranes as described in Chapter 3. The design peptides use amphiphilic structures, just as the phospholipids, to form well-ordered structures at the nano-scale by hydrophobic interactions. Several groups have utilized this approach including the groups of Samuel Stupp, Hiroshi Matsui, and Shugunag Zhang (Figure 6.3).

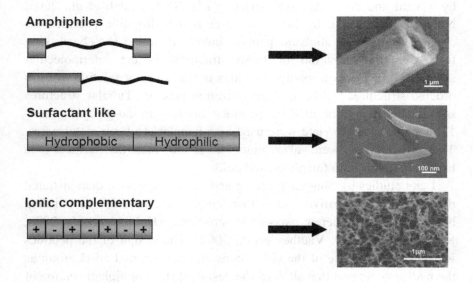

Figure 6.3: Formation of tubular and fibrilar structures by the self-assembly of short peptide by using hydrophobic or electrostatic interactions as driving force. Courtesy of Hiroshi Matsui and Shugunag Zhang. Modified from Reches and Gazit, 2006.

The work of Stupp and coworkers utilizes amphiphile molecules composed of a long hydrophobic aliphatic tail and a hydrophilic peptide chain head, similar to the structure of phospholipid molecules (Hartgerink *et al.*, 2001; Silva *et al.*, 2004). Yet, the peptide headgroups allow the integration of biological information into the nano-assemblies.

The suggested arrangement of the formed nano-fibrils is of a cylindrical micelle in which the peptide part is perpendicular to the fiber axis while the hydrophobic tail is placed in the fiber center and the hydrophobic head at the external part of the fiber. The researchers also use various modifications of the nanostructures such as cross-linking and mineralization. These were achieved by designing the peptide parts in such a way that they contain cysteine residues that allow interaction with the assemblies by thiol-metal interactions.

Peptide-based tubular structures can also self-assemble by linear peptides as was demonstrated with a family of bolaamphiphiles peptides by Matsui and co-workers (Banerjee *et al.*, 2003; Djalali *et al.*, 2003; Nuraje *et al.*, 2004). In these structures two hydrophilic peptides are conjugated through aliphatic peptide linker (Figure 6.3). The driving force for the formation of these structures is the intermolecular association of the hydrophobic moieties in the aqueous solution to form ordered structures similar to a micellization process. Tubular structures self-assembled by one of these peptides, bis(N-α-amido-glycylglycine)-1,7-heptane dicarboxylate were used as a template to form metal wires. These nanotubes were also immobilized to metallic surfaces via hydrogen bonds onto functionalised gold.

Later studies by Shugunag Zhang and co-workers have demonstrated the ability to use native, rather than conjugated, peptides as building blocks to form ordered nano-scale structures (Holmes *et al.*, 2000; Kisiday *et al.*, 2002; Vauthey *et al.*, 2002). The design of the peptides was based on the use of the side-chains and not external alkyl group as the hydrophobic part that allowed the design of the amphiphatic nature of the structures (Figure 6.3). The researchers designed a set of short peptides (7-8 amino acid) that used charged amino acid resides to build an hydrophilic head and aliphatic residues to construct a hydrophobic tail to form "surfactant-like" peptide molecules (Figure 6.3). Indeed these surfactant-like peptides undergo self-assembly to form nanotubes and nanovesicles. The surfactant-like peptide was also used as mimicry of biological membranes to extract and stabilize membrane proteins.

These nano-scale ordered peptides were also used to form macroscopic hydrogel assembly. Hydrogel scaffolds made of peptides with altering hydrophilic and hydrophobic amino acids were used as a template for tissue-cell attachment, extensive neurite outgrowth, and formation of active nerve connections (Holmes *et al.*, 2001).

6.5. Charge Complementary as a Driven Force for Self-Assembly

An earlier approach by Shuguang Zhang and co-workers, for the assembly of linear non-conjugated peptides, was based not only on hydrophobicity, but mainly on self-complementary ionic peptides (Altman *et al.*, 2000; Figure 6.3). These linear peptides are highly (>50%) charged with opposing charges using alternately basic and acidic amino acids, which is based on molecular organization of a segment within the zuotin yeast protein. Moreover the peptides are arranged to contain two distinct surfaces hydrophilic and hydrophobic. These peptides adopt β-sheet conformation and self-assemble into nanofibers. In some cases these nanofibers further form a hydrogel, which allows extensive neurite outgrowth and therefore can be used as a scaffold for tissue engineering.

6.6. Conjugation of Peptides for Self-Assembly

Another approach for the self-assembly of peptide structures at the nano-scale is based on the bis-conjugation of short peptide elements (Figure 6.4). In a very interesting work by Sandeep Verma and co-worker, short peptide elements were conjugated in a symmetric fashion to allow the formation of well-organized assemblies at the nano-scale (Madhavaiah and Verma, 2004, 2005). The conjugation of the short peptide elements allows the efficient assembly of structures that could be formed by the non-conjugated peptides. These include peptides that were derived from the prion protein that forms amyloids in the case of the deadly BSE syndrome (bovine spongiform encephalopathy or mad cow disease). The properties of the self-assembled amyloid nano-fibrils were already mentioned in Chapter 3 and will be described below. Clearly this self-organization of amyloid fibrils is a key process in biomolecular assembly that has ample technological applications.

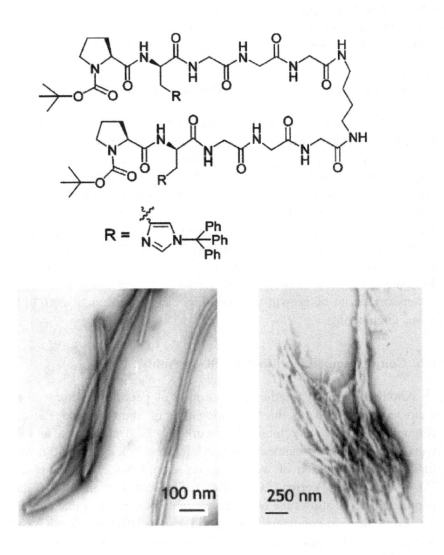

Figure 6.4: Formation of fibrilar nano-structures by the design of peptide conjugates. Courtesy of Sandeep Verma.

6.7. Aromatic Interactions for the Formation of Nanostructures

As mentioned above hydrophobicity is one of the central driving forces for the formation of peptide-based nanostructures. This self-organization is achieved in a similar manner as compared to the interaction between aliphatic moieties that lead to the formation of ordered structures. Another type of hydrophobic interaction that also provides order and directionality is the interaction between aromatic moieties.

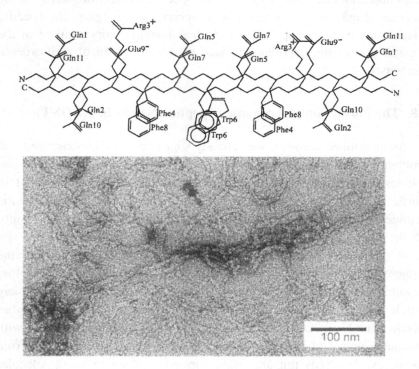

Figure 6.5: Formation of nano-tapes by the self-assembly of design β-sheet aromatic peptides. Courtesy of Sheena Radford and Neville Boden. Reprinted by permission from Macmillan Publishers Ltd: Nature (Aggeli *et al.*), copyright (1997).

The interaction between aromatic moieties is directional due to the exact restricted-geometry that is allowed between aromatic moieties (Hunter, 1993). As was mentioned in Chapter 3, these type of interactions were suggested to play a key role in the molecular recognition and self-assembly processes that lead to the formation of amyloid fibrils.

The use of specific aromatic interaction was already demonstrated in the 1990s to form nano-scale assemblies (Aggeli *et al.*, 1997). Sheena Radford and co-workers had designed a linear β-sheet peptide that contained aromatic moieties that made specific interactions between the moieties (Figure 6.5). As could be observed in the figure, the specific interactions between typtophan and phenylalanine pairs resulted in the organization of the peptides that lead to the formation of well-ordered nanotapes.

6.8. The Formation of Aromatic Dipeptide Nanotubes (ADNT)

As mentioned above, it was already suggested and demonstrated that aromatic interactions play a central role in the molecular recognition and self-assembly processes that lead to the formation of nano-scale amyloid fibrils. It was demonstrated by several groups that short aromatic peptides, as short as tetrapeptides, can form well-ordered amyloid nano-fibrils.

In the search for the shortest amyloidogenic peptide fragment, the diphenylalanine core motif of the Alzheimer's disease (AD) β-amyloid peptide was studied (Figure 6.6). Quite surprisingly this extremely simple peptide motif was found to form remarkably ordered nanotubes (Reches and Gazit, 2003; Song *et al.*, 2004). These hollow cylinders with internal diameter of about 20 nm and external diameter that ranges from 80 to few hundreds nm are being formed as discrete supramolecular assemblies with a very high persistence length. The peptides are being formed by either L- or D-amino-acid based peptides which have indistinguishable ultrastructural properties but differ in their proteolytic stability.

The applications of these nanotubes for nanotechnological applications were already demonstrated (Reches and Gazit, 2003; Song *et al.*, 2004; Yenimi *et al.*, 2005a,b) and will be further discussed in the next chapters.

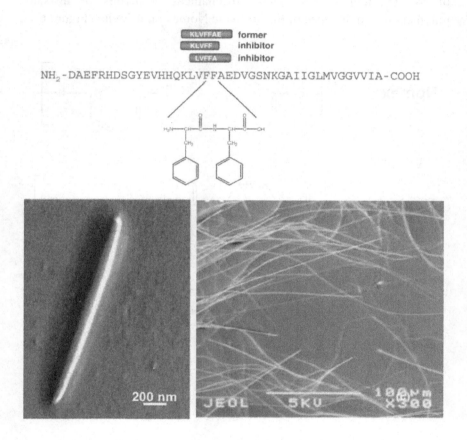

Figure 6.6: Formation of peptide nanotubes by the core recognition element of the β-amyloid polypeptide.

The aromatic dipeptide nanotubes are very attractive building blocks for nanotechnology as they show remarkable thermal and chemical stability (Adler-Abramovich *et al.*, 2006). The peptides nanotubes also show extraordinary mechanical strength. With a Young's modulus of 20 GPa, it appears to be the most rigid bio-inspired nanostructures. The extreme mechanical strength of the peptide nanotubes is suggested as the result of an extensive array of interactions between the aromatic moieties of the building blocks. These mechanical properties of aromatic polymers resemble those of the aromatic Nomex and Kevlar (Figure 6.7).

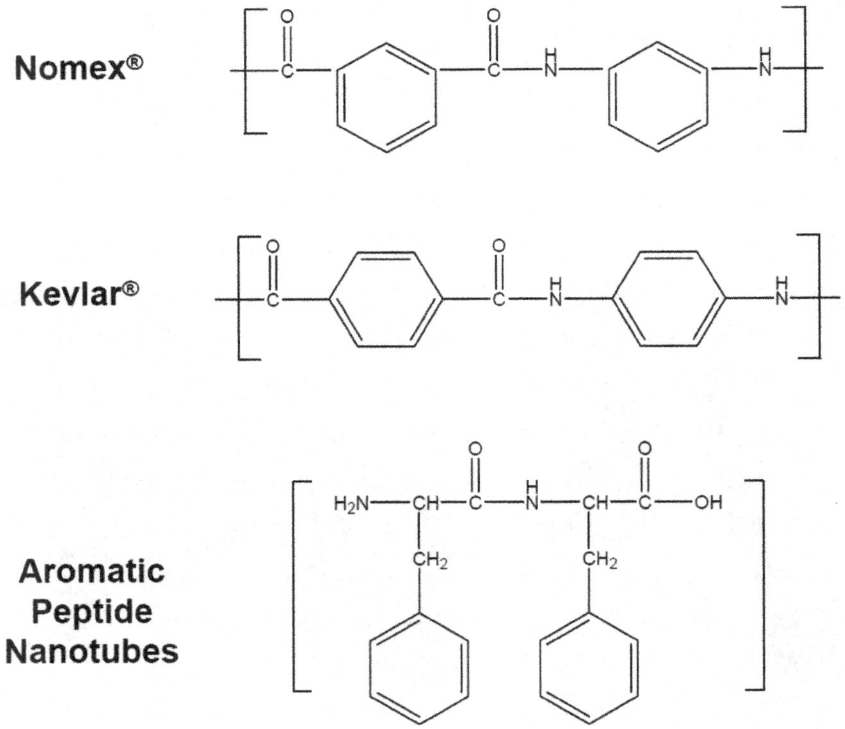

Figure 6.7: The chemical structure of the peptide nanotubes as compared to some of the strongest known aromatic nylons that are used for technological application.

These organic polymers are known to have very strong mechanical properties and they are used for many application. They are basically very similar to nylon, but have aromatic linkers instead of aliphatic ones.

The main difference between these polymers and the ADNT is the fact that the nanotubes are supramolecular assemblies that are formed by self-association of simple building blocks as compared to the high molecular-weight polymers.

6.9. The Formation of Spherical Nanostructures by Short Peptides

It was later revealed that diphenylglycine, a highly similar analogue and the simplest aromatic homo-dipeptide, does not form nanotubular structures but rather spherical nanometric assemblies (Figure 6.8; Reches and Gazit, 2004). The introduction of a thiol group into the diphenylalanine peptide also alters its assembly from tubular to spherical particles similar to those formed by diphenylglycine. The formation of either nanotubes or closed-cages by fundamentally similar peptides is consistent with a two-dimensional layer closure, as described in Chapter 2 both for carbon and inorganic nanotubes and their corresponding buckminsterfullerene and fullerene-like structures.

These spherical nanostructures may have various applications for drug delivery and imaging due to their very efficient assembly from such simple peptide building blocks and their regular morphology and biocompatibility. The ability to use the peptide nano-spheres for such applications is currently being explored.

6.10. Peptide Nucleic Acid (PNA)

The majority of this chapter is devoted to DNA structures that utilize the remarkable specific recognition of the nucleotide and to the peptides that serve as robust and very useful building blocks. A combination of both biomolecules is presented in peptide nucleic acids (PNA), a class of polymers with an amide backbone as of peptides but with nucleic acid bases as side chain (Nielsen, 1995). These molecules are synthesized using conventional peptide chemistry techniques as described above, but have much more advanced specific recognition properties as nucleic acids. Furthermore, the PNA polymers can form hetero-duplex assemblies with DNA or RNA and can be used for various molecular recognition applications.

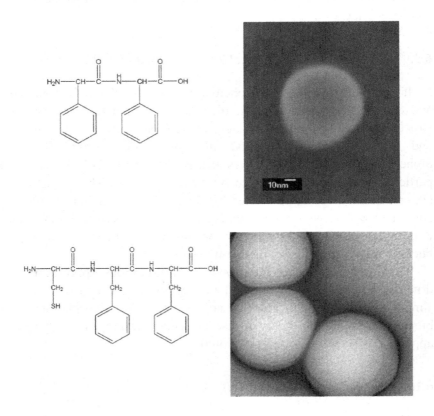

Figure 6.8: The formation of spherical nanostructures by the assembly of modified short aromatic peptides.

Chapter 7

Application of Biological Assemblies in Nanotechnology

In this chapter we will mainly consider bionanotechnology as defined in Chapter 1 (i.e., the use of biological and bio-inspired assemblies for nanotechnology). Biological assemblies have indeed already been successfully used in various nanotechnological applications. One of the most advanced uses of biological assemblies is the formation of metallic wires at the nano-scale. For this application various biomolecular assemblies were employed including DNA, proteins, and polypeptides. Some of the advancements in this direction include the use of molecular lithography and fabrication of coaxial nano-wires. Such nano-structure may have actual uses that are not related to biology but to molecular electronics.

7.1. The Use of S-Layer for Nanolithography

The formation of the nano-scale ordered S-layers was described in Chapter 3. These marvelous self-assembled nanostructures were demonstrated to be very useful for lithographic applications. Purified S-layer building blocks spontaneously re-assemble into well-ordered two-dimensional crystal under *in vitro* conditions. This property was used by Uwe Sleytr and co-workers to demonstrate that it is possible to recrystallize S-layer subunits on various substrates that are suitable for nanofabrication, such as silicon or silicone oxide wafers (Figure 7.1). Other studies have shown the ability to arrange the layers in lipid membrane or at air-water interface. S-layers have also been used as binding templates for well-organized arrangements of nanoparticles such as semiconductor and metallic dots (Gyorvary *et al.*, 2004) but also for biomolecules such as enzymes and antibodies. This type of assembly and molecular interaction allowed the application of the crystalline arrays as patterning media for nano-scale lithography (Figure 7.1).

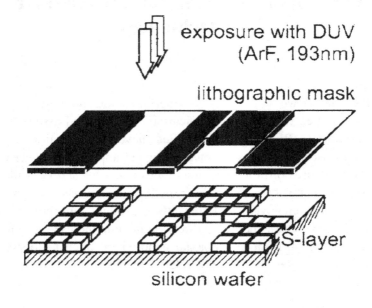

Figure 7.1: Schematic drawing of the patterning of an S-layer on a silicon wafer by DUV radiation using an ArF excimer laser. Courtesy of Uwe Sleytr.

It was demonstrated that it is possible to pattern S-layers using ultraviolet radiation (Figure 7.2). S-layers that were ordered on a silicon wafer are brought into direct contact with a photomask and exposed to ultraviolet irradiation. Such lithographic patterning of the layers, together with specific patterning of inorganic and biomolecules could allow the fabrication of a lab-on-a-chip. Other suggested applications of S-layer include the fabrication of ultrafiltration membranes with well-defined molecular weight cutoffs and immobilization matrices for the binding of monolayers of functional molecules in a geometrically well-defined way.

Figure 7.2: Scanning force microscopical image of a patterned S-layer on a silicon wafer. The bar represents a distance of 1000 nm. Courtesy of Uwe Sleytr.

7.2. The Use of DNA for Fabrication of Conductive Nanowires

The usefulness of DNA as a building material at the nano-scale, as was elegantly demonstrated by Nadrian Seeman, was already presented in Chapter 6. Yet, these marvelous DNA structures are rather inert. As we will discuss here, the specific properties of the DNA and its ability to specifically interact with various proteins have direct practical applications in nanotechnology. In this case DNA is being used as a nano-wire with specific recognition information embedded.

One of the earliest attempts and most elegant demonstrations of the use of DNA as bio-material for nanotechnology is the continuous work of Uri Sivan, Erez Braun, and co-workers (Figure 7.3). At the first stages of their nanotechnological applications of DNA, the controlled metal coating of DNA molecules was demonstrated (Braun *et al.*, 1988). The researchers had used a process of silver deposition and enhancement to form uniform metallic wire on a double stranded DNA template. The wires were of a dimension of about 100 nm and were proven to have ohmic conductivity.

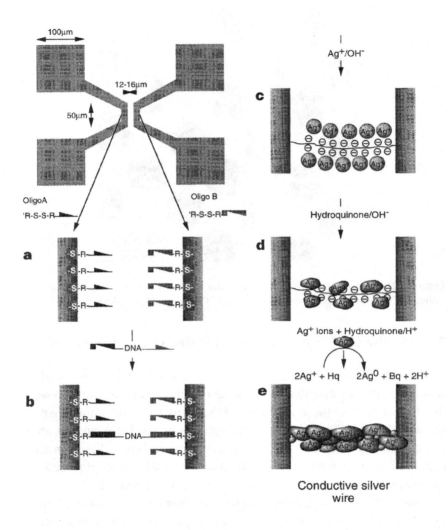

Figure 7.3: The metallization of DNA oligomers to form conductive nanowires. Single-strand DNA molecules were specifically targeted to surface-bound complementary DNA oligomers. The formed continuous DNA hybrid was further coated with silver to form conductive silver nano-wires. Reprinted by permission from Macmillan Publishers Ltd: Nature (Braun *et al.*), copyright (1998).

One of the great advantages of the DNA-based wires is the ability of the DNA to serve also as a carrier of address data that can help to direct the DNA strand into specific locations which is a key step for the achievement of self-assembled electronic circuits. Therefore, the same research groups also recently came one step further and demonstrated the process of molecular lithography of a DNA molecule (Keren *et al.*, 2002). The researchers had used RecA, a DNA protein that specifically binds unique sequences on the DNA strand (Figure 7.4).

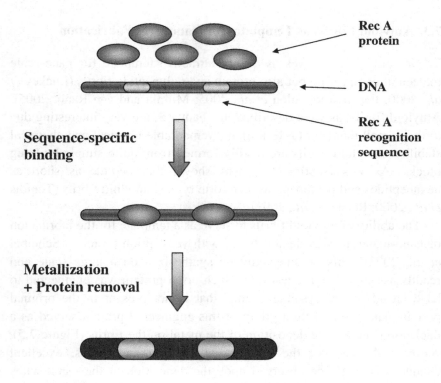

Figure 7.4: The use of the RecA protein for "molecular lithography". The RecA protein specifically binds to the unique sequence with the DNA and prevents the metallization of the specific region.

This design allowed the specific binding of the protein to the target DNA sequences and performed the metallization in a way that parts of the DNA molecule were protected from the chemical process. Thus, the method allowed a sequence specific and geometrically defined modification of the DNA molecule. The authors termed this process as molecular lithography. As was mentioned for conventional lithography in Chapter 2, the information encoded in the DNA molecules is equivalent to that of the masks used in conventional lithography, and the DNA-binding protein serves as the equivalent of a photoresist.

7.3. Amyloid Fibrils as Templates for Nanowires Fabrication

Not only DNA serves as a very attractive template for nano-scale molecular engineering but also protein and polypeptide fibrils (Luckey *et al.*, 2000; Papanikolopoulou *et al.*, 2005; Mitraki and van Raaij, 2005). Amyloid fibrils, as were described in Chapter 3, are very interesting due to their small diameter (7-10 nm) and remarkable physical and chemical stability. Amyloid fibrils are readily formed from quite simple building blocks. As was mentioned, it was shown that peptide as short as hexapeptides and pentapeptides can form typical amyloid fibrils (Tenidis *et al.*, 2000; Reches *et al.*, 2002).

The ability of amyloid fibrils to serve as a template for the fabrication of nano-materials was demonstrated with yeast prion proteins (Scheibel *et al.*, 2003). This protein could be synthesized on a large scale and readily assembled in the test tube. Such prion protein was engineered to have an additional cysteine residue that is not present in the original protein. The exposed thiol group of this engineered protein served as a nucleation site for the deposition of the metal on the fibrils (Figure 7.5). As with the case of the DNA-based metallic nano-wires, excellent conductivity could be observed upon the deposition of the metal wires between two electrodes.

The use of amyloid fibrils as a template for the fabrication of metal wires have a very practical prospect as very short peptide can form fibrils that are very similar to those formed by the full length proteins and polypeptides. Furthermore, it was shown that even a modified dipeptide can form such typical amyloid fibrils (Reches and Gazit, 2005). This could allow the large scale production of templates for the metallization and may provide a commercially viable solution for amyloid-based metallization.

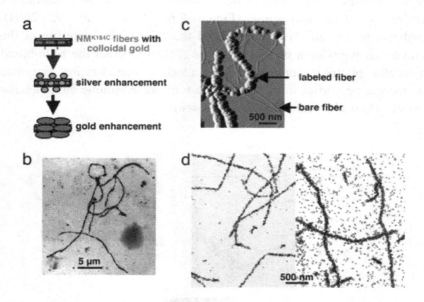

Figure 7.5: Metal coating of the nano-scale protein amyloid fibrils. a. Colloid gold binds to specific thiols within the amyloid building blocks, followed by silver and gold enhancements to achieve metal coated fibrils as observed by various microscopy techniques (**b-d**). Reproduced from Scheibel *et al.*, (2003) with permission. Copyright (2003) National Academy of Sciences, USA.

7.4. Metallization of Actin Filaments by Chemical Modification

Other biological fibrils at the nano-scale that were mentioned previously in this book are the actin filaments. These are self-assembled fibrils of 7 nm and have very uniform organization. Furthermore, as the crystal structure of the actin monomers is known, new functionalities could be rationally engineered and introduced into the fibrilar system. Thus these fibrils also serve as a very attractive platform for nano-scale engineering. One of the advantages of the actin filaments is the ability to control their polymerization and depolymerization by the presence of ATP nucleotides. It was indeed demonstrated by Itamar Willner and coworkers that re-polymerization of gold-coated actin filaments followed by the catalytic enlargement of the nanoparticles allow the formation of conductive metal nanowires (Figure 7.6; Patolsky *et al.*, 2004). Furthermore, the authors demonstrated the ATP-fuelled motility of the actin-based wires on a myosin interface. Taken together the actin-based molecular platform could combine selective metallization but also mechanical properties in the nano-scale forming a genuine biomolecular nanoelectromechanical (BioMEMS) system.

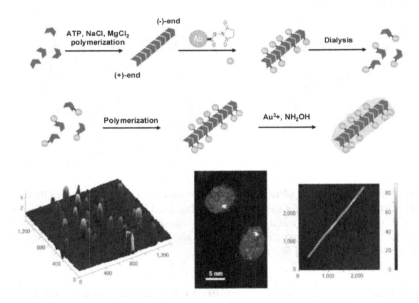

Figure 7.6: Metal coating of actin fibrils. The fibrils are chemically reacted to bind gold nano-particles, dialyzed, and re-polymerized with additional building blocks followed by gold enhancement. This results in continuous gold wires at the nano-scale. Reprinted by permission from Macmillan Publishers Ltd: Nature Materials (Patolsky *et al.*), copyright (2004). Courtesy of Itamar Willner.

7.5. The Use of Aromatic Peptide Nanotubes

Another group of bio-inspired materials that were used for the fabrication of metallic nano-wires are aromatic dipeptide nanotubes. These supramolecular assemblies are hollow with an inner diameter of few tens of nanometers. In the case of the tubes, the preferential entrance of metal ions into the lumen of the tubes allowed the formation of wires within the tubes rather than coating outside of the biolmolecule (Figure 7.6; Reches and Gazit, 2003). The use of biological proteolytic enzymes then allowed the degradation of the peptide coating and allowed us to obtain silver wire of a diameter of only 20 nm. This size of the metal wires is smaller that the one than can be achieved by conventional lithography. The remarkable rigidity of the nanotubes allows us to obtain elongated nanowires with very long persistence length (Figure 7.7).

As with the other biological materials that were described above, the peptide nanotubes could also be coated with an outer metallic layer. The combination of the deposition of metal with in the peptide tube but also in the outer layer allows the fabrication of coaxial metal wires in the nano-scale. Such coaxial wires should have several applications in molecular electronics.

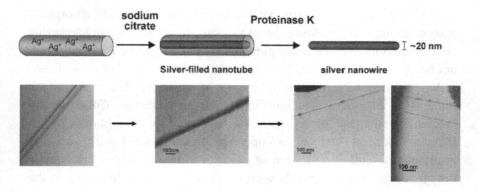

Figure 7.7: Formation of silver nanowires with peptide nanotubes templates. Ionic silver is reduced to metallic silver within the tube lumen. The peptide envelope is then removed by the use of enzymatic degradation. Modified from Reches and Gazit (2003).

7.6. Bacteriophages as Novel Biomaterials

Another very interesting building block for nanotechnology is bacteriophages, viruses that infect bacteria. In a pioneering work, Angela M. Belcher and coworkers were able to engineer phage particles for nanotechnology applications. The great advantage of the phage particles is their ability to express specific proteins on their surface in multiple copies. The utilization of these properties for the selection of high affinity peptide from a library was described in Chapter 5.

Belcher and coworkers described the use of short peptides that were selected for their high affinity to inorganic material to fabricate magnetic, semi-conductive and conductive nanowires (Lee *et al.*, 2002a; Mao *et al.*, 2004). Short peptides with affinity for various metals and semiconductors were displayed as coat proteins of the phage and were used for the fabrication of more complex assemblies with nano-scale order. A recent work by this group had demonstrated the usefulness of the phage-based nano-ordering strategy for the design and fabrication of lithium ion battery electrodes (Nam *et al.*, 2006).

7.7. The Use of Peptide Template for Biomineralization

Although most of the building blocks in biology are based on organic, carbon-dominated, building blocks such as proteins, nucleic acids, polysaccharides and phospholipids, the role of inorganic materials could not be ignored. The physical strength of skeletal tissues, bones, and teeth is based on the organization of inorganic materials. The most notable mineral is of course calcium in its different forms. Consequently, calcium uptake and organization has a central role in human health. Human disorders such as osteoporosis are associated with abnormal accumulation and organization of calcium in human bone. Yet, various organisms use other minerals such as silica as will be discussed in this chapter.

Bones clearly represent the most predominantly inorganic biological tissue. Bone contains about 30% organic material mostly in the form of collagen fibrils which provides the bone with strength and flexibility.

These are the same collagen protein fibrils that have a central role in other tissues such as skin and tendon, as were mentioned in Chapter 3. The remaining 70% of bone tissue is made up of inorganic materials. This calcium contains hydroxyapatite mineral, which is composed of various forms of calcium including calcium phosphate, calcium carbonate, calcium fluoride, calcium hydroxide and citrate. Most of the calcium in the bones exists in a crystalline rods with a thickness of 1-2 nm, width of 2-4 nm, and length of 20-40 nm.

Another important animal tissue that includes mostly inorganic material is dental tissue. The dentine layer of the tooth contains about 70% inorganic material, while the harder outer enamel coating contains about 98% inorganic material. As in the case of bone, the main inorganic constitute of the dental tissue is calcium-base hydroxyapatite mineral.

7.8. Production of Inorganic Composite Nanomaterial

Biological molecules serve in many cases as templates for the organization of macroscopic inorganic mater. This process is known as biomineralization. The formation of hard tissues in organisms such as those mentioned above (bone, teeth, and shells) is facilitated by the crystallization of inorganic minerals such as calcium carbonate using a biopolymeric template. The formed biocomposites are significantly different from the crystals that are formed by the inorganic material by itself with no template. Actually, the crystals may be "molded" into complicated macroscopic structures that have non-faceted crystal surfaces. All of this has of course been done at physiological temperature and pressure.

While calcium carbonate is the major interest in biomineralization that is related to human inorganic structures, it is worth mentioning that silica biomineralization is very common in marine organisms and allows to form remarkable patterning of silica at the nano-scale. One of the most vivid examples is that of diatom, micron-scale unicellular algae. These tiny creatures show astonishing organization of silica in their outer shells that have micron scale organization. These structures are compatible to silica structures that are being formed by lithography or at high temperature and pressure.

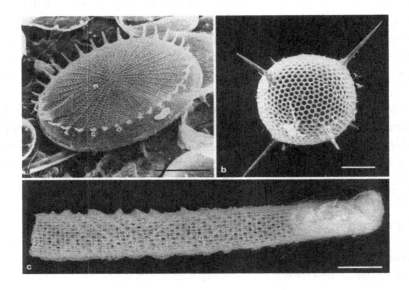

Figure 7.7: Marine organisms formed by directed organization of silica minerals. Courtesy of Carole Perry.

7.9. The Utilization of Biomineralization in Nanotechnology

There is a great interest in the utilization of biomineralization in nanotechnology. The ability to form complex inorganic structures such as artificial bones and teeth using biological templates that will direct inorganic organization may change the face of modern medicine. Much work is being done by examination various bioorganic templates that will allow crystallization of macroscopic structures.

Other work is directed toward the use of the diatoms principles for the fabrication of silicone at the nano-scale. This is especially intriguing as it may allow parallel fabrication of silicone-based structures as compared to the serial fabrication by lithographic methods. One example of such an approach is the use of a peptide fragment of silaffin-1, internal-repeats protein that mediates silica organization. The peptide fragment, which comprises the amino acid sequence of one repeat unit, can direct and affect the morphology of silica organization (Naik *et al.*, 2003).

Chapter 8

Medical and Other Applications of Bionanotechnology

One of the key future directions for nanobiotechnology and bionanotechnology is their medical use. As was mentioned briefly in the previous chapters, the use of nanostructures has already proven to be very useful in various applications, including drug delivery, ultrasensitive sensors, and the development of diagnostic tools. In this chapter we will first describe the application of nanostructures, biological and non-biological, as well as nanotechnology tools and techniques for various medical applications. The applications which will be described in this chapter are genuine and concrete and some of them are to be employed in the near future. In Chapter 9, we will describe some of the more futuristic applications of nano-science and nanotechnology in years to come such as tissue engineering and even *in situ* cellular DNA manipulation to treat genetic defects. While medical and human treatment applications represent a major part of theses activities, it is worth remembering that other applications are of high importance. This includes the uses of nanobiotechnology and bionanotechnology in agriculture, water technology (WaTech), cosmetics, and other applications as will be mentioned in the second part of the chapter.

8.1. The Use of Drug Nanocrystals for Improved Application

One of the most straightforward directions for nanotechnology to improve the effect of pharmaceutical products is to form nanocrystals of the desired drug. This is in contrast to the unordered powder or solidified powder as commonly used. The size of the nanocrystals is larger than that we described for inorganic crystals at the nano-scale and it is in the order of 200-500 nm. The nanocrystals are then administered as suspension in aqueous solution.

This method offers various advantages over simple formulation of the drugs. It offers much better solubility for hydrophobic drugs that are not water-soluble. It improves the bioavailability of the drug for absorption in the digestive track. Finally, it improves the stability of the drug to biological degradation in the body as well as the chemical and physical degradation as reflected by longer shelf-life.

Several nanodrug formulations were already approved by the FDA in the last few years. One of the early nanocrystal drugs was approved by the FDA is the immunosuppressant Rapamune® of Wyeth. The approved formulation in 2000, significantly improved the availability and pharmacokinetics of the drug. Current attempts are being made in the nano-formulation of large peptide drugs such as insulin and other hormones for oral administration.

8.2. The Use of Nano-Containers for Drug Delivery

A more complex way for the use of nanotechnology for drug-delivery is the use of heterologous nano-containers. As was briefly mentioned in our earlier discussion on the molecular basis of lipid vesicles self-assembly, one of the earliest applications of liposomes is the delivery of drugs to specific organs (Figure 8.1). The liposomes have many advantages due to their relative stability and ability to retain biological molecules with a low rate of leakage. Moreover they are composed of natural biomolecules and thus are recognized as "self" material by the human body. Their chemical functionality allows their decoration with various biological and chemical molecules for specific targeting of the vesicles by high affinity to specific ligands or receptors.

Such high affinity is crucial for the use of highly toxic drugs such chemotherapy agents that are necessary to delivered. The specific targeting of such liposomes exactly to the targeted tissue or organ allows us to use larger amounts of chemotherapy by reducing the overall side-effects by minimizing the exposure of the healthy tissue to the drug.

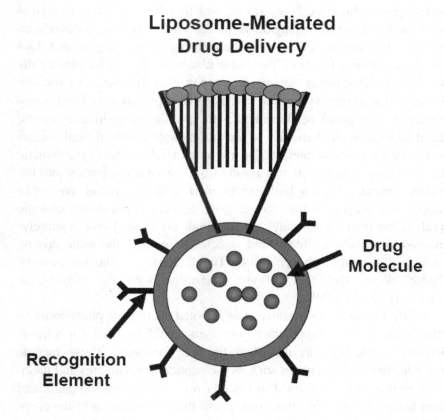

Liposome-Mediated Drug Delivery

Drug Molecule

Recognition Element

Figure 8.1: The use of phospholipid liposomes for drug delivery applications.
Liposome as closed-caged nano-containers can serve for various delivery applications.
The biocompatibility of the liposomes and ability to decorate them with biological and
chemical recognition elements allows their targeting to specific locations within the body.
The liposomes can be used to deliver not only drugs but also genetic material for gene-
therapy applications.

The encapsulation of the drug within the nano-container also affects
the absorption, distribution, and metabolism of the drug and can result in
better treatment protocols that are more friendly to the patient.

The liposomes can be used not only for the delivery of drugs but also
for the delivery of other materials such as DNA. This represents the
application of liposome technology for gene-therapy applications in each

normal gene which are being transferred to cells that carry a mutated non-functional one. This may be useful for various genetic diseases such as Cystic Fibrosis in which the newborn carries a mutated gene that does not allow a normal function. This could also be applicable for tumor cells that have mutated tumor suppressor genes that are the result of a somatic mutation due to the exposure to carcinogens. Liposomes, or other nano-containers that could be decorated with specific recognition elements, could be used as an alternative to the current application of viral vectors to deliver the genes of interest. The use of the viral vectors is problematic due to various side-effects that results from the viral integration into the patient genome. Thus a biologically inert delivery system may offer various advantages. In the case of gene delivery applications, cationic lipids rather than the naturally anionic lipids are used. These positively-charged phospholipids form lipid vesicles by exactly the same driving force as the natural phospholipids (Hart, 2005). Yet the net positive charge allows the high affinity interaction with the polyanionic negatively-charged DNA.

While the discussion above was devoted mainly to phospholipids, other nano-scale biocompatible containers could be used for various delivery applications as described for the liposomes. These include biomolecular-based vehicles such as the peptide nano-spheres and nano-vesicles that were mentioned in Chapter 6 as well as other organic and even inorganic structures that could allow the encapsulation of the cargo on the one hand but its desired release at the target on the other.

8.3. The Use of Inorganic Nanowires for Biological Detection

Other important tools in the development of ultra-sensitive devices are semiconductor nanowires, most importantly silicone nanowires. Charles M. Lieber and co-worker recently demonstrated a very sensitive use of silicone nanowires for biological applications (Patolsky *et al.*, 2004; Zheng *et al.*, 2005; Wang *et al.*, 2005). These one-dimensional nano-objects are modified with high-affinity antibodies (see Chapter 5) and their interaction could be detected at very high sensitivity using nanowire field effect transistors (Figure 8.2). The conductivity of the nanowires is so sensitive to the binding of the biological molecules that the binding of a single viral particle could be detected and recorded (Patolsky *et al.*, 2004). Such sensitivity has immense importance for many applications including early detection of various diseases, environmental and homeland security detection of biological and chemical agents, and the modulation of various nanodevices and nanomachines.

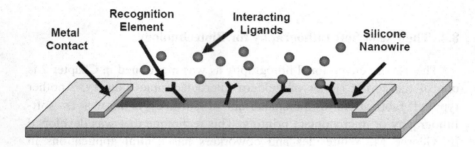

Figure 8.2: The use of silicone nanowires for the detection of biological interactions. The binding of charged ligands affects the current of either p-type or n-type silicone nanowires. The level of interaction could be monitored in real-time to provide information not only on the amount of the ligands but also on the kinetics of the association and dissociation.

In a later study, the authors demonstrated the electrical detection of biomarkers using multiplexed nanowire sensor arrays (Zheng *et al.*, 2005). Such a device allows the simultaneous monitoring of the interaction of large number of ligands. As in the case of the viral detection, the sensitivity was very high and sub-picogram concentrations of a biological marker could be detected by the array. The ability to monitor the electric signal as a function of time also allows the determination of the kinetics of the association and dissociation of the molecules. Thus, the nanowire multiarray serves as an alternative to the surface plasmon resonance (SPR) methods as mentioned in Chapter 4. As in the case of the SPR the nanowires allow the diction without the use of any external labeling. Yet, much higher sensitivity could be achieved. Nevertheless, it should be stressed that since the semiconductor nanowire are based on the gating of a transistor, the effect occurs only through the binding of charged ligands that can either increase or decrease the electrical current depending of the type of silicone nanowires. In the case of p-type nanowire (silicone that contains trivalent impurities such as boron or aluminum that are missing an electron in their outer electron shell), the conductance should decrease when the surface charge of ligand is positive and decrease when the surface charge is negative. The opposite is true for n-type silicone that contains atoms such as arsenic that contained an unpaired electron.

8.4. The Use of Soft Lithography for Biotechnology

The use of conventional lithography as was mentioned in Chapter 2 is one of the corner stones of modern microelectronic industry. Another type of fabrication of nanomaterials of surfaces is known as soft-lithography or microcontact printing. This technique that was developed by George M. Whitesides and coworkers has central applications in nanotechnology in general and nanobiotechnology in particular (Whitesides *et al.*, 2001; Shim *et al.*, 2003). The technique of soft-lithography is based on the use of a micro-patterned stamp that is made of a poly(dimethylsiloaxane) (PDMS). The elastic and transparent PDMS stamp is patterned using a fabricated master mold. The stamp is transparent not only to visible light by also to ultraviolet radiation down to 240 nm.

The PDMS stamp by itself is very useful for various biological applications. The physical properties of the stamp as mentioned above together with its excellent chemical resistance, thermal stability biocompetability make it an excellent system to pattern microfluidic devices with patterned water-filled microchannels. Such settings allow the probing of the properties of single cells under various experimental conditions. The response of cells and unicellular organisms to varied external signals, their changes under various chemical and biological conditions, their chemotaxis (response to gradient of chemical signals such as the attraction of the sperm to the egg), their cytometric (the chemical composition of the cells) and their differentiation could be studied at the single cell level (Li Jeon *et al.*, 2002; Wang and Ho, 2004; Wu *et al.*, 2004).

Another very interesting use of soft-lithography is its application for the printing of organic molecules including protein molecules. Under these conditions the stamp is inked with a solution that contains the organic molecules, the stamp is then rinsed and dried, following its printing of the surface. Such applications allowed the printing of various peptides and protein into nano-structured patterns (Kane *et al.*, 1999; Foley *et al.*, 2005). Such organization is useful for applications such as the fabrication of biosensors, the formation protein and DNA arrays, and the organization of biological signaling molecules for tissue engineering (as will be further elaborated in Chapter 10).

8.5. Contrast Agents by Nanomagnetic Materials

Another important application that should enjoy very important advancement due to nanotechnology developments in the field of MRI (Magnetic Resonance Imaging) contrast agents. MRI is an excellent imaging method in terms of safety and the ability for very accurate three dimensional tomography. Unlike X-ray based tomography techniques such as CT (computerized tomography) that is based on powerful X-ray electromagnetic radiation that can be harmful, the MRI techniques are based on the differential interaction of various biological entities with intense magnetic field. This represents a much safer application with long monitoring time. Indeed, MRI techniques are being used to follow dynamic processes with the body. Techniques such as function MRI (fMRI) allows to follow very complex processes such as cognition by continuous monitoring of the brain function upon the completion of various tasks. Yet, MRI like all magnetic resonance techniques is not very sensitive.

As this is a magnetic resonance technique, the imaging agents for MRI are ferromagnetic entities. This may include iron-based compounds, including oxides just like the magnetic material that we have on our computer hard disks, or metals such Gadolinium. The application of the ferromagnetic metals by themselves is limited due to questions of availability and safety and they are used as part of organic-based complexes. Nano-scale vehicles can allow the encapsulation of many ferromagnetic particles in dense packing to significantly enhance the sensitivity of the technique. Furthermore as with drug delivery, these agents can be specifically modified to allow specific targeting of the nano-vesicles. Targeting of the nano-carriers to specific disease-related receptors can allow the detection of events such as cancer.

8.6. Nanoagriculture

While most of the discussion regarding the utilization of nanotechnology is directed towards medical applications, it should be remembered that agriculture is also a very large and important field with important economical and social implications. Agriculture provides the food that we eat and the wood for furniture and construction. Furthermore, with the constant rising price of oil and natural gas and the depletion of the world reserves, in the future it might also be the source of energy as well as a source of starting materials for the chemical industry. Nanotechnology could actually provide many of the important diagnostics and therapeutic tool, that were initially directed for human use, also to agriculture. This is of course trivial in the case of animal science, as many of the therapeutic and diagnostic means that are being developed for human beings could be directly applied to domesticated animals. Just as vaccination and the use of antibiotics revolutionized the animal agriculture practice, so should future nanotechnological applications. Nanotechnology could also provide much advancement in plant growing agriculture. It should be noted that over the years, plant agriculture has become a very advanced scientific discipline and the production of agriculture products per given area is constantly increasing. The application of nanotechnology to this field includes the slow release of nutritional elements and minerals for nano-scale delivery systems. These systems could be designed in a way that would not just release the compounds in a constant pre-designed way, but could be nano-scale machine that will detect the exact situation of the plant and its external surroundings to provide the best optimize growth conditions.

8.7. Water Technology and Nanotechnology

Water technologies (WaTech) are quickly becoming a very important part of 21st century industry. Water technology includes various aspects of water desalination and treatment as well as the monitoring of water quality. The need for better technologies for the water industry is not only limited to arid areas with limited in water supply, but also to large industrial and urban areas that have to take care of the treatment of increasing volumes of sewage. Major companies such as General Electric and others had invested much effort to position themselves in this fast-growing field (see: http://www.gewater.com/).

The water technology field can have benefit from nanotechnological applications. Most of the water treatment facilities are based on the use of filters with nano-scale pores and inherent selectivity for water as compared to salt and pollutants. The design of new membranes that have better ultra-filtration performances, should lead both to the improvement of the quality as well as to the lowering of the cost of the desalination and water treatment. Bionanotechnology could have a special role in these efforts. As the materials to form the membrane should be biocompatible, hydrophilic, and non-toxic, biological assemblies such as nano-scale nanotubes could serve as nano-fluidic filtration elements in such applications.

Bionanotechnology could also be very useful in the treatment of the process of membrane biofouling in which microorganisms such as bacteria, fungi, and mold are accumulating on the surface of the membrane. The use of biological materials that are embedded in the filter and do not allow the growth of such undesirable microbial population is highly desirable. The anti-biofouling effect could be achieved by the nano-scale organization of physical and/or (bio)chemical agents that inhibit the growth of the bacteria. Alternatively slow release from nano-capsules may be used for the treatment.

Another important aspect in the application of biotechnology and nanotechnology in water technology is the monitoring of the quality of water. It is important to monitor both the level of biological and chemical agents, which could be added to the water in a malicious act. They could also be the by-product of the desalination process such as in the case of high levels of boron or radioactive salts when well-water rather than seawater is the initial source for the desalinated water.

8.8. Nanocosmetics

Another interesting application of nanotechnology which may represent a huge market is nanocosmetics. One of the desired applications by the cosmetics companies is to be able to supply the human skin with elements that are associated with younger appearance and physiological properties. This includes the introduction of various nourishing elements such as vitamins (e.g., the antioxidant lipophilic vitamin E that is limited for oral delivery), cofactors (as Q10), and even proteins (such as collagen fibrils).

The use of nano-carriers appears a very attractive mean to of applying such agents to the human skin. The cosmetic industry already utilizes liposomes as carriers of biomolecules to the skin. The use of liposome was already reported back in 1986 as part of the Capture® cosmetic formulation of Dior. Just as has been described for the drug delivery applications in the previous chapter, the liposomes serve as containers that can allow the delivery of a cargo to the desired part that is being topically treated.

Yet, the used of liposomes is limited only to hydrophilic water soluble substances as hydrophobic materials can actually affect the integrity of the liposomes whose self-assembly is driven by hydrophobic interactions. For these applications, peptide-based materials, especially those that use electrostatic interactions as the driving forces for assembly, may actually be much more attractive vehicles for such topical dermal delivery.

8.9. Solar Energy Applications

The use of bionanotechnology is also envisioned in the field of solar energy. The utilization of solar energy is always a very attractive field of research as the global energy consumption could by accommodated by solar energy if efficient devices were available. Unlike fossil fuel-based energy or atomic energy, solar energy is clean, inexpensive, and environmentally friendly. The recent increase in the price of oil and the declining amount of global reserves make the quest for alternative forms of energy (including geothermal and wind-based energy) very active.

The current photovoltaic solar cells are based on solid-state devices that allow the conversation of photons into a flow of electrons between doped semiconductor cells. One of the limiting factors in the utilization of solar energy is the efficiency of the commercial photocells which is about 10%. Due to large importance of solar energy much effort is being invested in nanotechnology in general to improve the efficiency of such devices to convert the sun photon flow into electric current. This includes new types of organic semiconductors and the use of titanium oxide nano-porous materials or quantum dots instead of semiconductor crystals. This is a very intriguing prospect that allows the improvement of research devices to over 30%.

Biology may actually offer much better improvement of the solar energy efficiency. As all readers will know, one of the essences of life is the conversion of solar energy into chemical energy via the process of photosynthesis. This process occurs in plants as well as in photosynthetic bacteria. The process was optimized throughout billions of years of evolution and the photosystem, the nanobiological assembly that facilitate the harvesting of the light and its conversation into chemical energy. The system had the incredible quantum efficiency of 1.0, which means that every photon that is harvested by the system is converted into excited state electron. There are several bionanotechnoloigcal attempts to integrate photosynthetic system into a solar cells. The use of proteins from photosynthetic bacteria that could withstand extreme conditions is highly desirable.

The group of Shugaung Zhang from MIT, who was mentioned earlier for his seminal work on the self-assembly of peptides, already demonstrated the proof of this principle of using biological photosystems assemblies in solar cells (Figure 8.3, Kiley *et al.*, 2005). The researchers had created a dense layer of light harvesting system molecules on another layer of organic semiconductor film which was itself on a metal electrode.

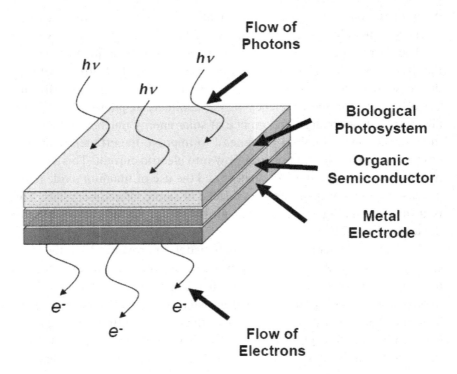

Figure 8.3: Biomolecular solar cell. As in silicone and other semiconductor-based photovoltaic cells, the biomolecular solar cell is based on the flow of electrons by the excitation of a semi-conductive material by electromagnetic radiation. The biological photosystem can significantly improve the efficiency of photon collection from the solar radiation.

Chapter 9

Future Prospects for Nanobiotechnology and Bionanotechnology

Nanobiotechnology and bionanotechnology are new scientific disciplines which are still in their infancy. While the prospects are enormous and unimaginable, as have been discussed throughout this book, many more directions are yet to be explored. In this chapter we will briefly discuss some of the directions envisioned for the years to come. Some of the directions are straightforward but others are visionary. We will begin with simple directions such as the marriage of genetic engineering and nanotechnology and the extension of the chemical horizons, while we will be concluding with some directions that may still be considered as "science fiction".

9.1. The Marriage of Molecular Biology and Nanotechnology

There is no dispute that molecular biology had a central role during the second half of the 20th century. While the previous century started with a true revolution in physics, its second half was dominated by the biological revolution. The ability to easily determine the DNA sequence of many organisms, the ability to clone DNA sequences followed by the expression and manipulation of protein and RNA molecules are just few examples. The budget of biological research in the USA and the rest of the world is continuously growing while the funding of other subjects slowly decreases. The ratio between the funding of NIH (National Institute Health) to NASA (National Aeronautics and Space Administration) is one of the indicators of this trend. In the 1960s, the funding for both agencies was about the same; as of 2006 the budget of

NIH is about three times higher than that of NASA (about 28 billion dollars for NIH and 11 billion dollars for NASA). Furthermore, many of the physical science and engineering departments, even including mechanical engineering or electrical engineering departments in traditionally engineering institute such as the Massachusetts Institute of Technology (MIT) and California Institute of Technology (CalTech), include biological research. Academic departments in areas such as chemistry or material science are of course even further associated with biological research. Therefore, as has happened in many other areas, the academic world already is been to integrate the physical and biological sciences. The same union could be observed also in the commercial world. As could be seen in Appendix II, many companies, large and small, conduct commercial endeavors that reflect the convergence of these two activities. Large semiconductor companies, such as Motorola, initiated activities with very dominant biological relevance.

9.2. The Engineering of Modified Biological Systems for the Assembly of Nanostructures

As discussed earlier in this book, the field of bionanotechnology is mainly based on the use of natural molecules, mostly DNA or proteins and small peptide fragments. One area that has not been fully explored is the combination of molecular biology and genetic engineering techniques together with the study of self-assembled biological building blocks. Molecular biology and genetic engineering may allow the integration of various biological functions into self-assembled protein structures. The formation of structures in which biological functions are integrated into nano-structures may be a central theme in the future. In principle it could allow the fabrication not only of well-structured assemblies in the nano-scale, but the formation of such structures that have functional units arranged in a nano-scale order. This could result in the ability to fabricate structures such nano-assembly lines in which various changes in a specific molecule could be carried out in nano-scale dimensions. We envision a situation in which the bionanotechnological engineer will be able to design a logical scheme for a systematic and reproducible fabrication of biological-based nanotechnological devices.

The field of nanobiotechnology on the other hand should also be significantly advanced by the marriage between concepts and tools from the physical sciences and molecular biology. Various techniques that were developed and are being developed for non-biological applications could be extremely useful for biological research. Many of the applications that are routinely being used in molecular biology could become much more efficient and robust through the use of nanotechnology. DNA manipulation and amplification, a very labor-intensive practice should be dramatically revolutionized by the use of microfluidics and "lab-on-a-chip" techniques. We could envision a situation in which the manipulation of a DNA sample could be automatically done on a chip in a sequential series of manipulations of a nano-fabricated chip.

9.3. Nanotechnology and Tissue Engineering

Tissue engineering is one of the most fascinating directions of modern biotechnology. The term "tissue engineering" that was coined by Robert S. Langer from MIT and refers to the ability to engineer and manufacture complete human tissues *in vitro*. The principle of the field of tissue engineering based on the ability to grow a network of cells, from patients themselves or donors, that will form functional tissues. This tissue could be used later for transplantation in human patients. The ability to conduct such manipulation should drastically change many medical practices and make some of the most severe disorders only part of the past. The ability for example to make an artificial pancreas in the laboratory will make Type I diabetes (juvenile diabetes) a completely curable disorder. Moreover, it may transform pancreatic cancer, one of the most devastating forms of cancer with extremely poor prognosis of only a few months of survival, into curable state. Other major tissues which might be engineered include of course the liver, heart, and kidney. All of these organs are currently repaired in the human body only by transplantation. Yet, there is a chronic lack of organs for transplantation and this process bears many complications due to the immune process. As stated above one of the major abilities of bionanotechnology is to use the patient's own cells as the mother cells for the engineered tissues. The resulting artificial tissue should therefore have the same immunological parameters as a "self" tissue and rejection should not be an issue anymore.

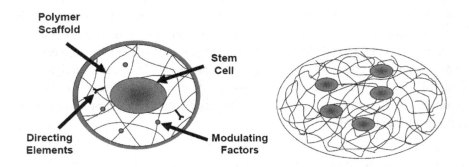

Figure 9.1: The use of a molecular scaffold for the growth of cells in a three dimensional matrix could serve as the basis for tissue engineering applications. The control of the tissue morphology and functioning by directing elements and modulating factors is highly desirable.

The basic premise of nanotechnology is to be able to fabricate a "smart matrix" with nano-scale order and signals for proliferation and differential of the cells (Figure 9.1). This matrix will serve as a scaffold for the positioning of stem-cells (either embryonic or adult stem cells) and providing them with the right conditions for the development of the engineered organs. The stems cells are unique in their ability to differentiate into cells with very different functional abilities. As all the genetic information exists in each and every cell, stem cells have the potential and the molecular information to become very distinct in their function, synthetic abilities, sensory abilities, and morphology. Some of the studies suggest the ability to control the differentiation of the cells by external electrical signals will achieve a way in which the process of organ formation is well-controlled.

The use of a polymeric scaffold for the manipulation of tissue was already demonstrated in the case of cartilage. In most elderly individuals the cartilage tissue is damaged. The dietary supplement of glucosamine and chondroitin sulfate may provide an improvement of the situation, but it far from being a complete solution. The cartilage tissue on the other hand is quite simple and no complex differentiation steps take place. Therefore, the engineering of cartilage tissue was one of the first targets to tackle. Cartilage cells were taken from a patient, grown in tissue culture conditions and then re-translated to the same individual. This prevented all forms of immunological rejection and allowed rapid functioning of the transplanted tissue. One of the major tasks of modern nanotechnology is undertake such practices in more complex organs such as the heart or liver in which the spatial organization of the cells and their differentiation according to a plan scheme is much more important.

In some other cases the ability of tissue regeneration may be a hidden function in the human body and nano-ordered materials could help to execute these abilities. This is especially true for injuries to the nervous system. It is argued that physical support, correct spatial engineering of cell proliferation signals, and biodegradability of engineered biomaterial could allow the recovery after spinal cord or optical nerve injuries if applied soon after the injury. If this will be achieved this could pave the way for transplantation of sensory organs, most importantly eyes. The source of the eyes could come from donors or even tissue engineered when the technology will allow.

9.4. Engineering of the Brain Tissue

Another major tissue engineering direction that is discussed less often, but of enormous medical importance, is the formation of brain tissue in the test tube. The formation of such structures is of course very fascinating as it can provide a cure to devastating diseases such as Alzheimer's disease and Parkinson's disease. As discussed in Chapter 4, these diseases are associated with the self-assembly of amyloid fibrils that results in the death of the surrounding tissue. These diseases represent a major medical problem. It is estimated that as of 2006 there are about 4 million Americans that are affected by Alzheimer's disease. A similar situation exists in the rest of the world. Since these diseases are associated with advanced age, it is assumed that as life expectancy increases, so will these amyloid-related diseases.

We currently know quite well how to grow individual nerve cells in a tissue culture, but the big challenge is to make it into a functional tissue. Here again, as discussed in the previous section, one of the directions is the organization of brain cells into a smart polymeric matrix in which intercellular axonal connections will be directed by spatial organization of signals.

Anther approach for the engineering of neural cells is a truly nanobiotechnological one (Kaul *et al.*, 2004). In this approach nerve cells are deposited on silicone substrates that are lithographically defined and the electrically activity of the cells is studied. The basic premise is to be able to form networks that will be complex enough to execute computational tasks. The final goal of course is to be able to create some kind of a biological computer of a silicone chip. This of course has great potential but also many very dangerous aspects as will be briefly discussed in the next chapter.

9.5. Making Artificial Biological Inorganic Composites

As the formation of functioning tissues represents true biological importance, it is no less important to be able to create functional bioinorganic composites. These include of course artificial bones and teeth. As was discussed earlier, biomineralization represents a major activity in biological systems as well as artificial nanobiotechnological applications. Therefore the directions which were explored and exploited by the in vitro systems may be further developed for the bioengineering of such tissues. Other directions may be different from the replication or mimicking of the calcium-based biomineralization. This could include nano-systems that are based on metals such as platinum and titanium that are currently used for the engineering of macroscopic transplanted objects.

Nanotechnology offers the ability to combine well-ordered structures at the nano-scale, the ordering of inorganic materials such as calcium and its derivatives, the ability to introduce biological signals, and materials of exceptional mechanical rigidity and chemical stability.

9.6. Nanobio Machines and Nanorobots

The ultimate goal of nanobiotechnology is the production of functional biological-relevant machines at the nano-scale. Such machines should be able to replace micro-surgery by using sub-micrometer precision of surgical procedures. One can envision the engineering of tiny robots that will have cellular dimensions of microns and actual surgical scalpels with sub-micromic dimensions. Such micro/nano-surgery robots could revolutionize the field of neurosurgery by allowing a very accurate manipulation of the damaged tissue as in brain tumors or damaged blood vessels without affecting the healthy tissue. Complicated brain surgery procedures, that often end in serious and irreversible complications, may be performed by such nano-robots. The idea of nano-robots was indeed the subject of science fiction as in the 1966 classics *Fantastic Voyage* by Isaac Asimov in which what we would call now a "nano-robot" was introduced into the body of one the characters and used to destroy a life-threatening blood clot in the brain.

Nanorobots could also be used for the treatment of disorders such as cancer on the cellular level. The key event in the transformation of normal cells into cancerous ones are mutations within the DNA sequence of affected individuals. For example most of human cancers are associated with mutation in the gene that encodes to the p53 tumor suppressor proteins. Correction of the function of this tumor suppressor, by protein fragments or small molecules, is indeed considered for future therapies. However other directions are possible. In a direction that still sounds like science fiction, nanomachines could detect the mutated DNA sequence, binding into the altered sequence, and then correct the DNA sequence in mutated cancerous cells by nano-scale resolution chemical correction.

Chapter 10

Concluding Remarks: The Prospects and Dangers of the Nanobiological Revolution

There is no more intellectually challenging task than to forecast the future. No one back in 1950s could even imagine the technological and cultural revolution that would be created by the introduction of personal computers and the internet. Nevertheless, without being able to grasp the full scale of future events, it was already very clear already by 1950 that the second half of the 20th century would be dominated by electronics and global telecommunication. When we stand now at the beginning of the 21st century attempting to have some insights into the future, it is very clear that the next fifty years will be dominated by a rapid and revolutionary advancement in both the fields of biology and nanotechnology. The fine details are yet to be revealed but the general framework can already be seen. Biology in the first decade of the 21st century is positioned approximately at the same place where physics was situated several decades or even a century ago. The general concepts are there, there is an explosion in the amount of experimental information, and there is even significant theoretical understanding of the processes. Most importantly, there is an immense interest and excitement regarding the importance of theses studies among the general public as well as among the world leaders and decisions makers.

One fascinating example was the press conference held simultaneously across the Atlantic by Bill Clinton and Tony Blair at the end the 1990s. During this conference they announced the successful completion of the human genome DNA sequencing project and discussed the implications of the project on the future of the citizens of the world. In spite of all the excitement and achievements, the pieces of this complicated jigsaw puzzle are not all in their right place and much more are needs to be learnt in the next fifty years. If we borrow the terminology that was used for physics a century ago, the world of biology is ready for the transition from classical biology into modern biology. Theories that are equivalent in their effect and magnitude to the theories of relativity and quantum mechanics are about to be discovered and formulated. The paradigm of physics research is a very good model for the comprehension of the process of the advancement of science in general and biology in particular. Physics underwent the progression from a discipline that was dominated by the accumulation of empirical data and its fitting to simple mathematical models into a scientific field that allows deep theoretical understanding of very complex processes with an incredibly strong prediction power. Many of the universal parameters, which were predicted theoretically by Einstein's theory of relativity, proved to be accurate only years later. Biology, as the central scientific discipline of the 21st century, is about to undergo the same conceptual revolution. The combination of biology together with nanotechnology should allow the application of physical and technological principles to the study of the complex living systems. As discussed throughout this book biology represents by far the most advanced collection of nano-devices, nano-assemblies and nano-machines.

When we discuss biology as a scientific field, it can be viewed from several different angles and levels. The very basic level is the discovery and formulation of the rules of the function of every living cell both in physiological and pathological states. There is a need to understand at the deepest level what makes cells live, divide, function, and die. What is the complete cascade of events that dictate the progression and fate of a living cell? What makes one cell different from the other? What makes the same cell function differently in one person as compared to another? What is going wrong when a normal cell is transformed into a malignant cell? Why is a newly divided cell different in an older person as compared to a young person or a newborn? This is only a very partial list of the different questions that fascinates every researcher and every person who is concerned about the meaning of life. As mentioned above, many of the details are known, but the outcome of the emerging fields of genomics and proteomics will provide the field with much deeper and more complete understanding. The recent completion of the human genome project is certainly a pivotal milestone in that direction. The next step will be the sequencing of the genomes of every individual and of any known species on earth. Such information will allow us to realize the similarities and differences between living individuals and will put the field of genetics in a completely new light. Nevertheless, the mere knowledge of the DNA sequences is not enough. A common analogy that is often made is that just looking at the list of parts of a Boeing 747 jet does not give a person, who never saw a plane, even the faintest hint about the way that this mammoth can fly. The next step in our understanding of the secrets of life is to study the main products of the genes, the proteins. This emerging field is called proteomics. The study of proteomics brings us one level higher. However, even this is not enough and we should obtain information about the map of the interaction between proteins, their structure and folding, and their ability to specifically interact with the DNA, by which they control the flow of information from the genome to the cells.

The next step after the understanding of the building blocks on the one hand and their way of assembly and function on the other will be the engineering of such activities. The prospects for the combination of the fabrication abilities and patterning as we learn from nanotechnology together with the understanding of the biological function can be revolutionary. Will we be able to design and manufacture living organisms as we design computers today? Can we manipulate living organisms as easily as we design integrated electronic circuits? These questions and others will be answered in years to come.

One of the most intriguing enigmas in biology is the function of the brain. In its essence, the study of the brain aims towards the understanding of the biochemical and molecular basis of thinking and reasoning. The brain is certainly the masterpiece of biological creation and the biggest mystery of all. Computers can calculate much faster than humans, and they can remember billions of phone numbers and financial transactions. However, no computer can think and no computer can make a moral decision. Above all, no computer can really perform a true recognition process that any young child can. Any four-year-old child can easily distinguish between a horse and a donkey. This simple task cannot be truly performed by a supercomputer with the most sophisticated software and hardware. There is much advancement in the field of artificial intelligence using newly mathematical tools like fuzzy logic and neural networks. These methods were developed with the aim of mimicking the computational process of the brain. However, with all the advancements in those fields, the most modern supercomputer cannot even compete with a young toddler or even a fruit fly. The reason for this is probably due to the fact that all these methods are missing some of the very fundamental concepts of brain function. In my mind, in the particular field of brain and cognition studies, we are still waiting for a complete new and revolutionary theory, in conceptual magnitude of the theories of relativity or quantum mechanics. Only such a revolutionary theory will be able to truly explain the process of thinking. On the way to breakthrough theory, nano-science and nanotechnology can play a central role. We would like to know about the molecular details of being mentally retarded or a genius? What is the net difference in the brain of a mouse before and after it knows the way through the maze in order to get to the piece of cheese? Moreover, will we be able to use technology, most interestingly nanotechnology, to fabricate artificial brain, a genuine an artificial intelligence device.

Many people ask themselves whether machines will ever have the ability to think. The answer is very clear and simple. There are machines that can think! The most elaborate one is called *Homo sapiens*. Unless we fully accept Descartes' body and mind dualism, there is no choice but to state that the brain is a machine that can think and that this thinking is done by the use of basic and finite chemical processes. The brain is a very elaborate machine, but it is just a machine that obeys the rules of chemistry and physics. There is no reason that such a machine will not eventually be built in a laboratory or later even in a mass-production assembly line. The bionanotechnological principles presented in this book allow to envision ways to make such complex machines by understanding the principles and practice of the assembly of a complex network of biological neurons into a silico-biological artificial device.

Finally, a few last words about the dangers of bionanotechnology and nanobiotechnology. Here, at the end of the book, we are come back to the parallels between biology and physics. The most dangerous outcome of modern physics was the atomic bomb. There has always been a great fear of a third world war in which atomic weapons will be used to destroy the world. Only a very fine network of checks and balances as well as the general understanding of the horrible outcome of the use of the bomb saved us from this devastation. The harmful potential of the nanotechnological revolution is no smaller than that of the atomic weapons. Actually it is much more dangerous. It is our duty to make sure that the nanotechnological tools will be used only for the best.

There's Plenty of Room at the Bottom: An Invitation to Enter a New Field of Physics – by Richard P. Feynman

The transcript of the original talk by Richard Feynman is reproduced with permission from Engineering & Science magazine, California Institute of Technology, Volume 23, Number 5, February 1960. It is presented to the readers as a vivid illustration the imagination and vision of one of the greatest scientific leaders of the 20th century.

Although the term "nanotechnology" is not mentioned in this lecture (as discussed in Chapter 2, it was suggested several years later), the address is certainly the cornerstone of modern nanotechnology. It is quite astonishing how well Feynman's remarks are still relevant almost half a century after his original address. As discussed previously, one of the most interesting things that is related to the readers of this book is Feynman' vision regarding the field of biology. Unlike many physicists of his time, Feynman understood how important biology is in the framework of nanotechnology.

There's Plenty of Room at the Bottom: An Invitation to Enter a New Field of Physics

Richard P. Feynman

I imagine experimental physicists must often look with envy at men like Kamerlingh Onnes, who discovered a field like low temperature, which seems to be bottomless and in which one can go down and down.

Such a man is then a leader and has some temporary monopoly in a scientific adventure. Percy Bridgman, in designing a way to obtain higher pressures, opened up another new field and was able to move into it and to lead us all along. The development of ever higher vacuum was a continuing development of the same kind.

I would like to describe a field, in which little has been done, but in which an enormous amount can be done in principle. This field is not quite the same as the others in that it will not tell us much of fundamental physics (in the sense of, "What are the strange particles?") but it is more like solid-state physics in the sense that it might tell us much of great interest about the strange phenomena that occur in complex situations. Furthermore, a point that is most important is that it would have an enormous number of technical applications.

What I want to talk about is the problem of manipulating and controlling things on a small scale.

As soon as I mention this, people tell me about miniaturization, and how far it has progressed today. They tell me about electric motors that are the size of the nail on your small finger. And there is a device on the market, they tell me, by which you can write the Lord's Prayer on the head of a pin. But that's nothing; that's the most primitive, halting step in the direction I intend to discuss. It is a staggeringly small world that is below. In the year 2000, when they look back at this age, they will wonder why it was not until the year 1960 that anybody began seriously to move in this direction.

Why cannot we write the entire 24 volumes of the Encyclopedia Brittanica on the head of a pin?

Let's see what would be involved. The head of a pin is a sixteenth of an inch across. If you magnify it by 25,000 diameters, the area of the head of the pin is then equal to the area of all the pages of the Encyclopaedia Brittanica. Therefore, all it is necessary to do is to reduce in size all the writing in the Encyclopaedia by 25,000 times. Is that possible? The resolving power of the eye is about 1/120 of an inch---that is roughly the diameter of one of the little dots on the fine half-tone reproductions in the Encyclopaedia. This, when you demagnify it by 25,000 times, is still 80 angstroms in diameter---32 atoms across, in an ordinary metal. In other words, one of those dots still would contain in its area 1,000 atoms. So, each dot can easily be adjusted in size as required by the photoengraving, and there is no question that there is enough room on the head of a pin to put all of the Encyclopaedia Brittanica.

Furthermore, it can be read if it is so written. Let's imagine that it is written in raised letters of metal; that is, where the black is in the Encyclopedia, we have raised letters of metal that are actually 1/25,000 of their ordinary size. How would we read it?

If we had something written in such a way, we could read it using techniques in common use today. (They will undoubtedly find a better way when we do actually have it written, but to make my point conservatively I shall just take techniques we know today.) We would press the metal into a plastic material and make a mold of it, then peel the plastic off very carefully, evaporate silica into the plastic to get a very thin film, then shadow it by evaporating gold at an angle against the silica so that all the little letters will appear clearly, dissolve the plastic away from the silica film, and then look through it with an electron microscope!

There is no question that if the thing were reduced by 25,000 times in the form of raised letters on the pin, it would be easy for us to read it today. Furthermore; there is no question that we would find it easy to make copies of the master; we would just need to press the same metal plate again into plastic and we would have another copy.

How do we write small?

The next question is: How do we write it? We have no standard technique to do this now. But let me argue that it is not as difficult as it first appears to be. We can reverse the lenses of the electron microscope in order to demagnify as well as magnify. A source of ions, sent through the microscope lenses in reverse, could be focused to a very small spot. We could write with that spot like we write in a TV cathode ray oscilloscope, by going across in lines, and having an adjustment which determines the amount of material which is going to be deposited as we scan in lines.

This method might be very slow because of space charge limitations. There will be more rapid methods. We could first make, perhaps by some photo process, a screen which has holes in it in the form of the letters. Then we would strike an arc behind the holes and draw metallic ions through the holes; then we could again use our system of lenses and make a small image in the form of ions, which would deposit the metal on the pin.

A simpler way might be this (though I am not sure it would work): We take light and, through an optical microscope running backwards, we focus it onto a very small photoelectric screen. Then electrons come away from the screen where the light is shining. These electrons are focused down in size by the electron microscope lenses to impinge directly upon the surface of the metal. Will such a beam etch away the metal if it is run long enough? I don't know. If it doesn't work for a metal surface, it must be possible to find some surface with which to coat the original pin so that, where the electrons bombard, a change is made which we could recognize later.

There is no intensity problem in these devices---not what you are used to in magnification, where you have to take a few electrons and spread them over a bigger and bigger screen; it is just the opposite. The light which we get from a page is concentrated onto a very small area so it is very intense. The few electrons which come from the photoelectric screen are demagnified down to a very tiny area so that, again, they are very intense. I don't know why this hasn't been done yet!

That's the Encyclopaedia Brittanica on the head of a pin, but let's consider all the books in the world. The Library of Congress has approximately 9 million volumes; the British Museum Library has 5 million volumes; there are also 5 million volumes in the National Library in France. Undoubtedly there are duplications, so let us say that there are some 24 million volumes of interest in the world.

What would happen if I print all this down at the scale we have been discussing? How much space would it take? It would take, of course, the area of about a million pinheads because, instead of there being just the 24 volumes of the Encyclopaedia, there are 24 million volumes. The million pinheads can be put in a square of a thousand pins on a side, or an area of about 3 square yards. That is to say, the silica replica with the paper-thin backing of plastic, with which we have made the copies, with all this information, is on an area of approximately the size of 35 pages of the Encyclopaedia. That is about half as many pages as there are in this magazine. All of the information which all of mankind has every recorded in books can be carried around in a pamphlet in your hand--- and not written in code, but a simple reproduction of the original pictures, engravings, and everything else on a small scale without loss of resolution.

What would our librarian at Caltech say, as she runs all over from one building to another, if I tell her that, ten years from now, all of the information that she is struggling to keep track of---120,000 volumes, stacked from the floor to the ceiling, drawers full of cards, storage rooms full of the older books---can be kept on just one library card! When the University of Brazil, for example, finds that their library is burned, we can send them a copy of every book in our library by striking off a copy from the master plate in a few hours and mailing it in an envelope no bigger or heavier than any other ordinary air mail letter.

Now, the name of this talk is "There is Plenty of Room at the Bottom"---not just "There is Room at the Bottom." What I have demonstrated is that there is room---that you can decrease the size of things in a practical way. I now want to show that there is plenty of room. I will not now discuss how we are going to do it, but only what is possible in principle---in other words, what is possible according to the laws of physics. I am not inventing anti-gravity, which is possible someday only if the laws are not what we think. I am telling you what could be done if the laws are what we think; we are not doing it simply because we haven't yet gotten around to it.

Information on a small scale

Suppose that, instead of trying to reproduce the pictures and all the information directly in its present form, we write only the information content in a code of dots and dashes, or something like that, to represent the various letters. Each letter represents six or seven "bits" of information; that is, you need only about six or seven dots or dashes for each letter. Now, instead of writing everything, as I did before, on the surface of the head of a pin, I am going to use the interior of the material as well.

Let us represent a dot by a small spot of one metal, the next dash, by an adjacent spot of another metal, and so on. Suppose, to be conservative, that a bit of information is going to require a little cube of atoms 5 times 5 times 5---that is 125 atoms. Perhaps we need a hundred and some odd atoms to make sure that the information is not lost through diffusion, or through some other process.

I have estimated how many letters there are in the Encyclopaedia, and I have assumed that each of my 24 million books is as big as an Encyclopaedia volume, and have calculated, then, how many bits of information there are (10^{15}). For each bit I allow 100 atoms. And it turns out that all of the information that man has carefully accumulated in all the books in the world can be written in this form in a cube of material one two-hundredth of an inch wide---which is the barest piece of dust that can be made out by the human eye. So there is plenty of room at the bottom! Don't tell me about microfilm!

This fact---that enormous amounts of information can be carried in an exceedingly small space---is, of course, well known to the biologists, and resolves the mystery which existed before we understood all this clearly, of how it could be that, in the tiniest cell, all of the information for the organization of a complex creature such as ourselves can be stored. All this information---whether we have brown eyes, or whether we think at all, or that in the embryo the jawbone should first develop with a little hole in the side so that later a nerve can grow through it---all this information is contained in a very tiny fraction of the cell in the form of long-chain DNA molecules in which approximately 50 atoms are used for one bit of information about the cell.

Better electron microscopes

If I have written in a code, with 5 times 5 times 5 atoms to a bit, the question is: How could I read it today? The electron microscope is not quite good enough, with the greatest care and effort, it can only resolve about 10 angstroms. I would like to try and impress upon you while I am talking about all of these things on a small scale, the importance of improving the electron microscope by a hundred times. It is not impossible; it is not against the laws of diffraction of the electron. The wave length of the electron in such a microscope is only 1/20 of an angstrom. So it should be possible to see the individual atoms. What good would it be to see individual atoms distinctly?

We have friends in other fields---in biology, for instance. We physicists often look at them and say, "You know the reason you fellows are making so little progress?" (Actually I don't know any field where they are making more rapid progress than they are in biology today.) "You should use more mathematics, like we do." They could answer us--- but they're polite, so I'll answer for them: "What you should do in order for us to make more rapid progress is to make the electron microscope 100 times better."

What are the most central and fundamental problems of biology today? They are questions like: What is the sequence of bases in the DNA? What happens when you have a mutation? How is the base order in the DNA connected to the order of amino acids in the protein? What is the structure of the RNA; is it single-chain or double-chain, and how is it related in its order of bases to the DNA? What is the organization of the microsomes? How are proteins synthesized? Where does the RNA go? How does it sit? Where do the proteins sit? Where do the amino acids go in? In photosynthesis, where is the chlorophyll; how is it arranged; where are the carotenoids involved in this thing? What is the system of the conversion of light into chemical energy?

It is very easy to answer many of these fundamental biological questions; you just look at the thing! You will see the order of bases in the chain; you will see the structure of the microsome. Unfortunately, the present microscope sees at a scale which is just a bit too crude. Make the microscope one hundred times more powerful, and many problems of biology would be made very much easier. I exaggerate, of course, but the biologists would surely be very thankful to you---and they would prefer that to the criticism that they should use more mathematics.

The theory of chemical processes today is based on theoretical physics. In this sense, physics supplies the foundation of chemistry. But chemistry also has analysis. If you have a strange substance and you want to know what it is, you go through a long and complicated process of chemical analysis. You can analyze almost anything today, so I am a little late with my idea. But if the physicists wanted to, they could also dig under the chemists in the problem of chemical analysis. It would be very easy to make an analysis of any complicated chemical substance; all one would have to do would be to look at it and see where the atoms are. The only trouble is that the electron microscope is one hundred times too poor. (Later, I would like to ask the question: Can the physicists do something about the third problem of chemistry---namely, synthesis? Is there a physical way to synthesize any chemical substance?

The reason the electron microscope is so poor is that the f- value of the lenses is only 1 part to 1,000; you don't have a big enough numerical aperture. And I know that there are theorems which prove that it is impossible, with axially symmetrical stationary field lenses, to produce an f-value any bigger than so and so; and therefore the resolving power at the present time is at its theoretical maximum. But in every theorem there are assumptions. Why must the field be symmetrical? I put this out as a challenge: Is there no way to make the electron microscope more powerful?

The marvelous biological system

The biological example of writing information on a small scale has inspired me to think of something that should be possible. Biology is not simply writing information; it is doing something about it. A biological system can be exceedingly small. Many of the cells are very tiny, but they are very active; they manufacture various substances; they walk around; they wiggle; and they do all kinds of marvelous things---all on a very small scale. Also, they store information. Consider the possibility that we too can make a thing very small which does what we want---that we can manufacture an object that maneuvers at that level!

There may even be an economic point to this business of making things very small. Let me remind you of some of the problems of computing machines. In computers we have to store an enormous amount of information. The kind of writing that I was mentioning before, in which I had everything down as a distribution of metal, is permanent.

Much more interesting to a computer is a way of writing, erasing, and writing something else. (This is usually because we don't want to waste the material on which we have just written. Yet if we could write it in a very small space, it wouldn't make any difference; it could just be thrown away after it was read. It doesn't cost very much for the material).

Miniaturizing the computer

I don't know how to do this on a small scale in a practical way, but I do know that computing machines are very large; they fill rooms. Why can't we make them very small, make them of little wires, little elements ---and by little, I mean little. For instance, the wires should be 10 or 100 atoms in diameter, and the circuits should be a few thousand angstroms across. Everybody who has analyzed the logical theory of computers has come to the conclusion that the possibilities of computers are very interesting---if they could be made to be more complicated by several orders of magnitude. If they had millions of times as many elements, they could make judgments. They would have time to calculate what is the best way to make the calculation that they are about to make. They could select the method of analysis which, from their experience, is better than the one that we would give to them. And in many other ways, they would have new qualitative features.

If I look at your face I immediately recognize that I have seen it before. (Actually, my friends will say I have chosen an unfortunate example here for the subject of this illustration. At least I recognize that it is a man and not an apple.) Yet there is no machine which, with that speed, can take a picture of a face and say even that it is a man; and much less that it is the same man that you showed it before---unless it is exactly the same picture. If the face is changed; if I am closer to the face; if I am further from the face; if the light changes---I recognize it anyway. Now, this little computer I carry in my head is easily able to do that. The computers that we build are not able to do that. The number of elements in this bone box of mine are enormously greater than the number of elements in our "wonderful" computers. But our mechanical computers are too big; the elements in this box are microscopic. I want to make some that are submicroscopic.

If we wanted to make a computer that had all these marvelous extra qualitative abilities, we would have to make it, perhaps, the size of the Pentagon. This has several disadvantages. First, it requires too much material; there may not be enough germanium in the world for all the transistors which would have to be put into this enormous thing. There is also the problem of heat generation and power consumption; TVA would be needed to run the computer. But an even more practical difficulty is that the computer would be limited to a certain speed. Because of its large size, there is finite time required to get the information from one place to another. The information cannot go any faster than the speed of light---so, ultimately, when our computers get faster and faster and more and more elaborate, we will have to make them smaller and smaller.

But there is plenty of room to make them smaller. There is nothing that I can see in the physical laws that says the computer elements cannot be made enormously smaller than they are now. In fact, there may be certain advantages.

Miniaturization by evaporation

How can we make such a device? What kind of manufacturing processes would we use? One possibility we might consider, since we have talked about writing by putting atoms down in a certain arrangement, would be to evaporate the material, then evaporate the insulator next to it. Then, for the next layer, evaporate another position of a wire, another insulator, and so on. So, you simply evaporate until you have a block of stuff which has the elements---coils and condensers, transistors and so on---of exceedingly fine dimensions.

But I would like to discuss, just for amusement, that there are other possibilities. Why can't we manufacture these small computers somewhat like we manufacture the big ones? Why can't we drill holes, cut things, solder things, stamp things out, mold different shapes all at an infinitesimal level? What are the limitations as to how small a thing has to be before you can no longer mold it? How many times when you are working on something frustratingly tiny like your wife's wrist watch, have you said to yourself, "If I could only train an ant to do this!" What I would like to suggest is the possibility of training an ant to train a mite to do this. What are the possibilities of small but movable machines? They may or may not be useful, but they surely would be fun to make.

Consider any machine---for example, an automobile---and ask about the problems of making an infinitesimal machine like it. Suppose, in the particular design of the automobile, we need a certain precision of the parts; we need an accuracy, let's suppose, of 4/10,000 of an inch. If things are more inaccurate than that in the shape of the cylinder and so on, it isn't going to work very well. If I make the thing too small, I have to worry about the size of the atoms; I can't make a circle of "balls" so to speak, if the circle is too small. So, if I make the error, corresponding to 4/10,000 of an inch, correspond to an error of 10 atoms, it turns out that I can reduce the dimensions of an automobile 4,000 times, approximately ---so that it is 1 mm. across. Obviously, if you redesign the car so that it would work with a much larger tolerance, which is not at all impossible, then you could make a much smaller device.

It is interesting to consider what the problems are in such small machines. Firstly, with parts stressed to the same degree, the forces go as the area you are reducing, so that things like weight and inertia are of relatively no importance. The strength of material, in other words, is very much greater in proportion. The stresses and expansion of the flywheel from centrifugal force, for example, would be the same proportion only if the rotational speed is increased in the same proportion as we decrease the size. On the other hand, the metals that we use have a grain structure, and this would be very annoying at small scale because the material is not homogeneous. Plastics and glass and things of this amorphous nature are very much more homogeneous, and so we would have to make our machines out of such materials.

There are problems associated with the electrical part of the system ---with the copper wires and the magnetic parts. The magnetic properties on a very small scale are not the same as on a large scale; there is the "domain" problem involved. A big magnet made of millions of domains can only be made on a small scale with one domain. The electrical equipment won't simply be scaled down; it has to be redesigned. But I can see no reason why it can't be redesigned to work again.

Problems of lubrication

Lubrication involves some interesting points. The effective viscosity of oil would be higher and higher in proportion as we went down (and if we increase the speed as much as we can). If we don't increase the speed so much, and change from oil to kerosene or some other fluid, the problem is not so bad. But actually we may not have to lubricate at all! We have a lot of extra force. Let the bearings run dry; they won't run hot because the heat escapes away from such a small device very, very rapidly.

This rapid heat loss would prevent the gasoline from exploding, so an internal combustion engine is impossible. Other chemical reactions, liberating energy when cold, can be used. Probably an external supply of electrical power would be most convenient for such small machines.

What would be the utility of such machines? Who knows? Of course, a small automobile would only be useful for the mites to drive around in, and I suppose our Christian interests don't go that far. However, we did note the possibility of the manufacture of small elements for computers in completely automatic factories, containing lathes and other machine tools at the very small level. The small lathe would not have to be exactly like our big lathe. I leave to your imagination the improvement of the design to take full advantage of the properties of things on a small scale, and in such a way that the fully automatic aspect would be easiest to manage.

A friend of mine (Albert R. Hibbs) suggests a very interesting possibility for relatively small machines. He says that, although it is a very wild idea, it would be interesting in surgery if you could swallow the surgeon. You put the mechanical surgeon inside the blood vessel and it goes into the heart and "looks" around. (Of course the information has to be fed out.) It finds out which valve is the faulty one and takes a little knife and slices it out. Other small machines might be permanently incorporated in the body to assist some inadequately-functioning organ.

Now comes the interesting question: How do we make such a tiny mechanism? I leave that to you. However, let me suggest one weird possibility. You know, in the atomic energy plants they have materials and machines that they can't handle directly because they have become radioactive. To unscrew nuts and put on bolts and so on, they have a set of master and slave hands, so that by operating a set of levers here, you control the "hands" there, and can turn them this way and that so you can handle things quite nicely.

Most of these devices are actually made rather simply, in that there is a particular cable, like a marionette string, that goes directly from the controls to the "hands." But, of course, things also have been made using servo motors, so that the connection between the one thing and the other is electrical rather than mechanical. When you turn the levers, they turn a servo motor, and it changes the electrical currents in the wires, which repositions a motor at the other end.

Now, I want to build much the same device---a master-slave system which operates electrically. But I want the slaves to be made especially carefully by modern large-scale machinists so that they are one-fourth the scale of the "hands" that you ordinarily maneuver. So you have a scheme by which you can do things at one-quarter scale anyway---the little servo motors with little hands play with little nuts and bolts; they drill little holes; they are four times smaller. Aha! So I manufacture a quarter-size lathe; I manufacture quarter-size tools; and I make, at the one-quarter scale, still another set of hands again relatively one-quarter size! This is one-sixteenth size, from my point of view. And after I finish doing this I wire directly from my large-scale system, through transformers perhaps, to the one-sixteenth-size servo motors. Thus I can now manipulate the one-sixteenth size hands.

Well, you get the principle from there on. It is rather a difficult program, but it is a possibility. You might say that one can go much farther in one step than from one to four. Of course, this has all to be designed very carefully and it is not necessary simply to make it like hands. If you thought of it very carefully, you could probably arrive at a much better system for doing such things.

If you work through a pantograph, even today, you can get much more than a factor of four in even one step. But you can't work directly through a pantograph which makes a smaller pantograph which then makes a smaller pantograph---because of the looseness of the holes and the irregularities of construction. The end of the pantograph wiggles with a relatively greater irregularity than the irregularity with which you move your hands. In going down this scale, I would find the end of the pantograph on the end of the pantograph on the end of the pantograph shaking so badly that it wasn't doing anything sensible at all.

At each stage, it is necessary to improve the precision of the apparatus. If, for instance, having made a small lathe with a pantograph, we find its lead screw irregular---more irregular than the large-scale one---we could lap the lead screw against breakable nuts that you can reverse in the usual way back and forth until this lead screw is, at its scale, as accurate as our original lead screws, at our scale.

We can make flats by rubbing unflat surfaces in triplicates together---in three pairs---and the flats then become flatter than the thing you started with. Thus, it is not impossible to improve precision on a small scale by the correct operations. So, when we build this stuff, it is necessary at each step to improve the accuracy of the equipment by working for awhile down there, making accurate lead screws, Johansen blocks, and all the other materials which we use in accurate machine work at the higher level. We have to stop at each level and manufacture all the stuff to go to the next level---a very long and very difficult program. Perhaps you can figure a better way than that to get down to small scale more rapidly.

Yet, after all this, you have just got one little baby lathe four thousand times smaller than usual. But we were thinking of making an enormous computer, which we were going to build by drilling holes on this lathe to make little washers for the computer. How many washers can you manufacture on this one lathe?

A hundred tiny hands

When I make my first set of slave "hands" at one-fourth scale, I am going to make ten sets. I make ten sets of "hands," and I wire them to my original levers so they each do exactly the same thing at the same time in parallel. Now, when I am making my new devices one-quarter again as small, I let each one manufacture ten copies, so that I would have a hundred "hands" at the 1/16th size.

Where am I going to put the million lathes that I am going to have?

Why, there is nothing to it; the volume is much less than that of even one full-scale lathe. For instance, if I made a billion little lathes, each 1/4000 of the scale of a regular lathe, there are plenty of materials and space available because in the billion little ones there is less than 2 percent of the materials in one big lathe.

It doesn't cost anything for materials, you see. So I want to build a billion tiny factories, models of each other, which are manufacturing simultaneously, drilling holes, stamping parts, and so on.

As we go down in size, there are a number of interesting problems that arise. All things do not simply scale down in proportion. There is the problem that materials stick together by the molecular (Van der Waals) attractions. It would be like this: After you have made a part and you unscrew the nut from a bolt, it isn't going to fall down because the gravity isn't appreciable; it would even be hard to get it off the bolt. It would be like those old movies of a man with his hands full of molasses, trying to get rid of a glass of water. There will be several problems of this nature that we will have to be ready to design for.

Rearranging the atoms

But I am not afraid to consider the final question as to whether, ultimately---in the great future---we can arrange the atoms the way we want; the very atoms, all the way down! What would happen if we could arrange the atoms one by one the way we want them (within reason, of course; you can't put them so that they are chemically unstable, for example).

Up to now, we have been content to dig in the ground to find minerals. We heat them and we do things on a large scale with them, and we hope to get a pure substance with just so much impurity, and so on. But we must always accept some atomic arrangement that nature gives us. We haven't got anything, say, with a "checkerboard" arrangement, with the impurity atoms exactly arranged 1,000 angstroms apart, or in some other particular pattern.

What could we do with layered structures with just the right layers? What would the properties of materials be if we could really arrange the atoms the way we want them? They would be very interesting to investigate theoretically. I can't see exactly what would happen, but I can hardly doubt that when we have some control of the arrangement of things on a small scale we will get an enormously greater range of possible properties that substances can have, and of different things that we can do.

Consider, for example, a piece of material in which we make little coils and condensers (or their solid state analogs) 1,000 or 10,000 angstroms in a circuit, one right next to the other, over a large area, with little antennas sticking out at the other end---a whole series of circuits. Is it possible, for example, to emit light from a whole set of antennas, like we emit radio waves from an organized set of antennas to beam the radio programs to Europe? The same thing would be to beam the light out in a definite direction with very high intensity. (Perhaps such a beam is not very useful technically or economically.)

I have thought about some of the problems of building electric circuits on a small scale, and the problem of resistance is serious. If you build a corresponding circuit on a small scale, its natural frequency goes up, since the wave length goes down as the scale; but the skin depth only decreases with the square root of the scale ratio, and so resistive problems are of increasing difficulty. Possibly we can beat resistance through the use of superconductivity if the frequency is not too high, or by other tricks.

Atoms in a small world

When we get to the very, very small world---say circuits of seven atoms---we have a lot of new things that would happen that represent completely new opportunities for design. Atoms on a small scale behave like nothing on a large scale, for they satisfy the laws of quantum mechanics. So, as we go down and fiddle around with the atoms down there, we are working with different laws, and we can expect to do different things. We can manufacture in different ways. We can use, not just circuits, but some system involving the quantized energy levels, or the interactions of quantized spins, etc.

Another thing we will notice is that, if we go down far enough, all of our devices can be mass produced so that they are absolutely perfect copies of one another. We cannot build two large machines so that the dimensions are exactly the same. But if your machine is only 100 atoms high, you only have to get it correct to one-half of one percent to make sure the other machine is exactly the same size---namely, 100 atoms high!

At the atomic level, we have new kinds of forces and new kinds of possibilities, new kinds of effects. The problems of manufacture and reproduction of materials will be quite different. I am, as I said, inspired by the biological phenomena in which chemical forces are used in repetitious fashion to produce all kinds of weird effects (one of which is the author).

The principles of physics, as far as I can see, do not speak against the possibility of maneuvering things atom by atom. It is not an attempt to violate any laws; it is something, in principle, that can be done; but in practice, it has not been done because we are too big.

Ultimately, we can do chemical synthesis. A chemist comes to us and says, "Look, I want a molecule that has the atoms arranged thus and so; make me that molecule." The chemist does a mysterious thing when he wants to make a molecule. He sees that it has got that ring, so he mixes this and that, and he shakes it, and he fiddles around. And, at the end of a difficult process, he usually does succeed in synthesizing what he wants. By the time I get my devices working, so that we can do it by physics, he will have figured out how to synthesize absolutely anything, so that this will really be useless.

But it is interesting that it would be, in principle, possible (I think) for a physicist to synthesize any chemical substance that the chemist writes down. Give the orders and the physicist synthesizes it. How? Put the atoms down where the chemist says, and so you make the substance. The problems of chemistry and biology can be greatly helped if our ability to see what we are doing, and to do things on an atomic level, is ultimately developed---a development which I think cannot be avoided.

Now, you might say, "Who should do this and why should they do it?" Well, I pointed out a few of the economic applications, but I know that the reason that you would do it might be just for fun. But have some fun! Let's have a competition between laboratories. Let one laboratory make a tiny motor which it sends to another lab which sends it back with a thing that fits inside the shaft of the first motor.

High school competition

Just for the fun of it, and in order to get kids interested in this field, I would propose that someone who has some contact with the high schools think of making some kind of high school competition. After all, we haven't even started in this field, and even the kids can write smaller than has ever been written before. They could have competition in high schools. The Los Angeles high school could send a pin to the Venice high school on which it says, "How's this?" They get the pin back, and in the dot of the "i" it says, "Not so hot."

Perhaps this doesn't excite you to do it, and only economics will do so. Then I want to do something; but I can't do it at the present moment, because I haven't prepared the ground. It is my intention to offer a prize of $1,000 to the first guy who can take the information on the page of a book and put it on an area 1/25,000 smaller in linear scale in such manner that it can be read by an electron microscope.

And I want to offer another prize---if I can figure out how to phrase it so that I don't get into a mess of arguments about definitions---of another $1,000 to the first guy who makes an operating electric motor---a rotating electric motor which can be controlled from the outside and, not counting the lead-in wires, is only 1/64 inch cube.

I do not expect that such prizes will have to wait very long for claimants.

Appendix B

List of Bionanotechnological and Nanobiotechnological Companies

The field of nanobiotechnology and bionanotechnology is a very dynamic one. As in other hi-tech commercial activities it combines well-established companies (such as mega-size chemical manufactures) as well as younger companies and newborn start-ups. Here is a list of active companies as of 2006:

3DM: (http://www.puramatrix.com/): Production self-assembled biomaterials with nano-scale orders for the application of tissue engineering and repair.

3DM Inc.
P.O. Box 425025
Cambridge, MA 02142, USA
E-mail: contact@puramatrix.com

Acrongenomics: (http://www.acrongen.com/): "A research & development nanobiotechnology company pioneering uniquely advanced nanomolecular diagnostics for the Life Sciences Industry".

Acrongenomics, Inc.
38A Poseidonos Ave.
17455 Alimos
Athens, Greece
E-mail: info@acrongen.com

Ademtech: (http://www.ademtech.com/): The production of calibrated magnetic emulsions for in-vitro diagnosis and life sciences application.

Ademtech
Parc scientifique Unitec 1
4 allée du doyen Georges Brus
33600 Pessac, France
E-mail: ademtech@ademtech.com

Adnavance Technologies: (http://www.adnavance.com/) Technology applications that use the conductivity properties of both normal and novel, metallic, forms of DNA, aimed at large biosensor/molecular diagnostics and energy generation/management markets.

Adnavance Technologies Inc.
860 - 410, 22nd St. East
Saskatoon SK S7K 5T6, Canada
E-mail: info@adnavance.com

Alnis BioSciences: (http://www.alnis.com/): The production of "engineered nanoscopic hydrogels, or NanoGels, comprised of polymers, bioactives, and targeting molecules. These NanoGels are intended to selectively target pathological tissues, treat the disease, and afterwards simply disintegrate, returning to the individual components from which they were formed".

Alnis BioSciences, Inc.
626 Bancroft Way #3B
Berkeley, CA 94710, USA
E-mail: alnis@alnis.com

Advance Nanotech: (http://www.advancenanotech.com): The acquisition and commercialization of nanotechnology in the areas of electronics, biopharma and materials.

Advance Naotech Inc.
600 Lexington Ave.
New York, NY 10022, USA
E-mail: info@advancenanotech.com

Agilent Technologies: (http://www.agilent.com): (spin-off Hewlett-Packard Company). Advance measurement tools for life-science and medical use.

Agilent Technologies, Inc.
395 Page Mill Rd.
Palo Alto, CA 94306, USA
E-mail: Contact_Us@agilent.com

Artificial Cell Technologies: (http://www.artificialcelltech.com/): The production of artificial cells as an artificial bionanomachine, using a polypeptide engineering technology platform.

Artificial Cell Technologies, Inc.
4 Research Dr.
Shelton, CT 06484, USA

Ambi Limited: (http://www.ambri.com/): "pioneering the integration of Biotechnology, Nanotechnology and Electronics with a major focus in the human medical diagnostics market".

Ambri Ltd.
126 Greville St.
Chatswood NSW 2067, Australia
E-mail: communications@ambri.com

Asklêpios   BioPharmaceutical: (http://www.askbio.com/):
"Biotechnology company engaged in the development and delivery of
novel protein and cellular based therapies through design of proprietary
Biological Nano Particles ('BNP') ".

Asklêpios BioPharmaceutical Inc.
510 Meadowmount Village Circle
Chapel Hill, NC 27517, USA
E-mail: info@askbio.com

BioCrystal: (http://www.biocrystal.com/):   Production   of
nanocrystalline fluorescent markers for cell and intracellular detection
and analysis.

BioCrystal, Ltd.
575 McCorkle Boulevard
Westerville, OH 43082, USA
Email: info@biocrystal.com

BioForce Nanosciences: (http://www.bioforcenano.com/):   The
development of ultra-miniaturized array technologies for solid-phase,
high-throughput biomolecular analysis.

BioForce Nanosciences, Inc.
1615 Golden Aspen Dr.
Ames, IA 50010, USA
E-mail: info@bioforcenano.com

BioNano International Singapore: (http://www.bionano.com.sg/):
The development of various advanced sensor devices through the use of
nanotechnology including the application of carbon nanotubes.

BioNano International Singapore Pte Ltd
8 Prince George's Park
NUS Business Incubator, Singapore 118407
E-mail: info@bionano.com.sg

Degussa Advanced Nanomaterials: (http://www.advancednano
materials.com/): The production of nanostructured materials and their
dispersions.

Degussa AG
Postcode 1040-004
Rodenbacher Chaussee 4
63457 Hanau-Wolfgang, Germany
E-mail: adnano@degussa.com

Dendritech: (http://www.dendritech.com/): "Dendritech is a specialty
polymer company focused on the development of dendritic,
hyperbranched and tailor-made polymer products".

Dendritech, Inc.
3110 Schuette Dr.
Midland, MI 48642, USA
E-mail: scheibert@dendritech.com

Dendritic Nanotechnologies: (http://dnanotech.com/): "DNT is the
world's leading developer and provider of advanced dendritic polymers,
a breakthrough enabling technology that provides the vehicle — the
targeting and delivery mechanisms — for a vast array of diagnostic and
therapeutics solutions — *in vitro* and *in vivo*".

Dendritic NanoTechnologies Inc
2625 Denison Dr.
Mount Pleasant, MI 48858, USA
E-mail: info@dnanotech.com

Evident   Technologies: (http://www.evidenttech.com/):   The
production of quantum dots of various compositions and surface
modifications for life-science research applications.

Evident Technologies
216 River St.
Troy, NY 12180, USA
E-mail: info@evidenttech.com

Flamel Technologies: http://www.flamel.com/: Flamel Technologies is a drug delivery company with expertise in polymer chemistry. "Delivery systems for small molecule and protein drugs, providing tailored solutions to the biotech and pharmaceutical industries for optimized controlled-release delivery of drugs".

Flamel Technologies
33 avenue du Dr. Georges Levy
69693 Venissieux Cedex, France
E-mail: info@flamel.com

GeneFluidics: (http://www.genefluidics.com/): "By integrating novel bionano and microfluidic technologies, the Company's revolutionary platform will enable complex tests that are currently performed by skilled technicians in a laboratory to be performed by anyone, anywhere".

GeneFluidics
2540 Corporate Place
Monterey Park, CA 91754
E-mail: info@GeneFluidics.com

General Electric Healthcare (formally Amersham Biosciences Corp). (http://www.gehealthcare.com/): A major player in the field on molecular medicine and the utilization of novel nanotechnology applications.

GE Healthcare
800 Centennial Ave.
Piscataway, NJ 08855, USA

ImaRx Therapeutics: (http://www.imarx.com/): "The world leader in NanoInvasive™ medicine". ImaRx Therapeutics is utilizing groundbreaking medical technologies to develop products with the potential to revolutionize the treatment of strokes, and vascular disease. ImaRx faces near-term commercial opportunities in addition to a deep pipeline of development stage compounds.

ImaRx Therapeutics, Inc.
1635 East 18th St.
Tucson, AZ 85719, USA
E-mail: imarx@imarx.com

JR Nanotech: (http://www.jrnanotech.com/). "Using Nano-silver technology to solve age old healthcare problems".

JR Nanotech
145 Chase Rd.
London N14 4JP, UK
E-mail: info@jrnanotech.com

Kereos: http://www.kereos.com/: Target therapeutic and molecular imaging.

Kereos, Inc.
4041 Forest Park Ave.
Saint Louis, MO 63108, USA
E-mail: info@kereos.com

Keystone Nano: http://www.keystonenano.com/: The production of molecular dots (MD) for diagnosis and treatment.

Ketstone Nano, Inc.
158 Round Hill Rd.
Boalsburg, PA 16827, USA
E-mail: info@keystonenano.com

KnowmTech: (http://www.knowmtech.com/): the development of nanotechnology-based neural networks.

KnowmTech, LLC
117 Byrn Mawr Dr. SE
Albuquerque, NM 87106, USA

Medical Murray: (http://www.medicalmurray.com/): "Medical Murray has extensive experience in the injection molding and development of nanoscaled plastic parts".

Medical Murray
400 N Rand Rd.
North Barrington
IL 60010, USA

Mobious Genomics: (http://www.mobious.com/): The production of diagnostic and research tools by unifying biological molecular systems and artificial nanostructures.

Mobious Genomics Ltd.
University Innovation Centre
University of Exeter
Rennes Road
Exeter EX4 4QJ, UK
E-mail: contact@mobious.com

Nanobac Life Sciences: (http://www.nanobaclabs.com/): Research is establishing the pathogenic role of Calcifying Nano-Particles (CNPs) or "Nanobacteria" in human disease.

Nanobac Life Sciences, Inc.
2727 W. Martin Luther King Blvd.
Tampa, FL 33607, USA
E-mail: info@nanobaclabs.com

NanoBio® Corporation: (http://www.nanobio.com/): "Bio-pharmaceutical company developing and commercializing therapies based on proprietary nanoemulsion technology for the treatment of viral and fungal skin infections and vaginal infections".

NanoBio® Corporation
Mail: P.O. Box 8110
Ann Arbor, MI 48107, USA
E-mail: marketing@nanobio.com

NanoBioMagnetics: (http://www.nanobmi.com/):   "A   nano-biomaterials   company   engaged   in   the   development   and commercialization of magnetically responsive nanoparticle technologies (MNP) for a range of human health applications".

NanoBioMagnetics, Inc.
124 N. Bryant Ave.
Edmond, OK 73034
E-mail: nanobio@nanobmi.com

Nanocopoeia: (http://www.nanocopeia.com/): "A drug formulation company bringing a proprietary ElectroNanoSpray™ technology for producing nanoparticles to the pharmaceutical and medical device industries".

Nanocopoeia, Inc.
1246 West University Ave.
St. Paul, MN 55104, USA
E-mail: info@nanocopoeia.com

Nanogen: (http://www.nanogen.com/): The use of nanotechnology to advance the application of DNA array technology for diagnosis applications.

Nanogen Corporate Headquarters
10398 Pacific Center Court
San Diego, CA 92121, USA
E-mail: technicalassistance@nanogen.com

NanoHorizons: (http://nanohorizons.com/): "Develop and produce state of the art nanotechnology products and processes for near term biotechnology, pharmaceutical, chemical and microelectronic applications".

NanoHorizons Inc.
Technology Center, Suite 208
200 Innovation Blvd.
State College, PA 16803, USA
E-mail: info@NanoHorizons.com

NanoInk: (http://www.nanoink.net/): The use of Dip Pen Nanolithography™ method for the nano-scale printed of various molecules, including biomolecules, for nanotechnological applications.

NanoInk, Inc.
1335 Randolph St.
Chicago, IL 60607
E-mail: info@nanoink.net

Nanolayers: (http://www.nanolayers.com/): The combination of organics electronic and nanotechnology for various applications including the production of biosensors.

Nanolayers
11 Alfassi St.
Jerusalem 92302, Israel
E-mail: ben@nanolayers.com

Nanomix: (http://www.nano.com/): The development of ultrasensitive detections systems based on carbon nanotubes combined with proprietary chemistries.

Nanomix, Inc.
5980 Horton Street
Emeryville, CA 94608, USA
E-mail: info@nano.com

Nanosphere: (http://www.nanosphere-inc.com/): "Harnessing the unique properties of matter at the nano-scale in order to commercialize the proprietary chemistry (nanoparticle probes, bio-barcode™

technology and automated instruments) that is the foundation of its platform".

Nanosphere, Inc.
4088 Commercial Avenue
Northbrook, IL 60062
E-mail: info@nanosphere.us

Nanotherapeutics: (http://www.nanotherapeutics.com/): "Nanotherapeutics has developed two proprietary drug delivery technologies, Nanodry™ and Nanocoat™, which produce improved pharmaceuticals at high yield that are substantially more versatile and cost-effective than current technologies".

Nanotherapeutics, Inc.
12085 Research Drive
Alachua, FL 32615, USA
E-mail: info@nanotherapeutics.com

Nanoxis: (http://www.nanoxis.se/):   "A   nano-biotechnology company determined to provide generic and optimized tools for membrane protein identification and analysis. Our unique concept is built upon new nanofabrication technologies using soft polymer and fluid-state liquid crystalline materials".

Nanoxis A.B.
MC2 Building A at Chalmers
Kemivägen 9
SE-412 96 Gothenburg, Sweden
E-mail: info@nanoxis.com

Orla Protein Technologies: (http://www.orlaproteins.com/): The interface between material science and biology to study the surface of proteins and produce "high density protein arrays, protein coated particles for therapeutics, bioseparations, surfaces for the attachment and modulation of cells, implantable devices and nanoscale diagnostics".

Orla Protein Technologies Ltd
Nanotechnology Centre
Newcastle Upon Tyne
NE1 7RU, UK
E-mail: enquiries@orlaproteins.com

Platypus Technologies: (http://www.platypustech.com/): "Platypus Technologies, LLC develops, produces and markets nanotechnologies for the life-sciences. We produce a variety of nano-structured surfaces for use in research and are also developing a range of products that derive from a proprietary platform technology using liquid crystals for the rapid detection of molecular interactions".

Platypus Technologies, LLC
5520 Nobel Dr., Suite 100
Madison, WI 53711
E-mail: info@platypustech.com.

Protiveris: (http://www.protiveris.com/): The use of Microcantilevers technology, an ultrasensitive nano-mechanical surface stress and mass sensors for the diction of molecular interaction events.

Protiveris Inc.
15010 Broschart Road
Rockville, MD 20850, USA

Quantum Dot Corp: (http://www.qdots.com): The production of quantum dots of various composition and surface modifications for life-science research applications.

Quantum Dot Corp.
26118 Research Road
Hayward, CA 94545, USA

Schoeller Textil: (http://www.nano-sphere.ch): The production of self-cleaning clothes. "The NanoSphere® finishing technology is based on the self-cleaning principle and is nearly a perfect copy of nature".

Schoeller Textil AG
Bahnhofstrasse 17
CH- 9475 Sevelen
E-mail: info@schoeller-textiles.com

Starpharma: http://www.starpharma.com: The of dendrimer nanotechnology for "the development of high-value dendrimer nanodrugs to address unmet market needs".

Starpharma Holdings Ltd
75 Commercial Road
Melbourne VIC 3004, Australia
E-mail: info@starpharma.com

Triton BioSystems: http://www.tritonbiosystems.com/: The development of novel anti-cancer therapy based on magnetic nano-materials and magnetic field energy.

Triton BioSystems, Inc.
200 Turnpike Road
Chelmsford, MA 01824, USA
E-mail: Inquire@TritonBioSystems.com

Velbionanotech: http://www.velbionanotech.com/: "VBN is dedicated to furthering the art, science and practice of nanotechnology, where Biology, Chemistry, Physics and mechanical design converge Molecular scale".

Velbionanotech
City Point,Infantry Road,
Bangalore-560 001, India
E-mail: info@velbionanotech.com

Appendix C

Glossary

Actin filament: A two-stranded helical polymer with a diameter of about 7 nm. Serves as a major protein component of the cytoskeleton and striated muscles. Also known as *microfilaments* (also see *myosin* and *polymer*).

Affinity: The interaction between two or more chemical entities as reflected by their dissociation constant.

Amino acid: The building blocks of proteins and peptides polymers. A typical amino acid is composed of a chiral carbon linked to an amino group, carboxyl group, and a functional side chain (also see *peptide*, *polymer*, *protein*).

Amphiphilic or **Amphipathic:** A chemical compound that have both hydrophilic and hydrophobic nature (from the Greek *amphis*: both, and *philia*: love).

Amyloid fibril: An ordered protein or peptide fibril with a diameter of 7-10 nm that is associated with human disease such as Alzheimer's disease and Type II diabetes.

Angstrom (Å): One tenth of a *nanometer*, 10^{-10} meter.

Antibody or **Immunoglobulin:** A protein complex that binds specific chemical entities ("antigen") at high affinity. Antibodies serve as a major tool of the immune system to combat bacterial or viral infections (also see *antigen*).

Antigen: A chemical compound that stimulates an immune response, especially the production of antibodies (also see *antibody*).

Archea or **Archeabacteria:** A branch of prokaryotes that is different form bacteria in many parameters and resembles eukaryotes in some aspects. Believed to be a third primary biological kingdom besides bacteria and eukaryotes.

Aromatic compound: A compound that contained one or more cyclic planar carbon moieties that includes shared resonant electrons. Aromatic compounds spontaneously organized in geometrically-restricted stacking interactions.

Avidin: A glycoprotein that has a very strong affinity for biotin with dissociation constant about to 10^{-15} M^{-1}, the strongest known for non-covalent interactions (also see *covalent interactions*).

Atomic force microscopy (AFM): A type of scanning probe microscope, with very high-resolution. AFM serves for imaging, measuring and manipulating matter at the nano-scale.

Bacterium (*pl* Bacteria)**:** A unicellular prokaryote microorganisms that is classified as lower form of life.

Bacteriophage: A virus which infects bacteria and manipulate in them (from bacteria and Greek *phagein*: to eat).

Biomineralization: A process by which organisms produce minerals, often to harden or stiffen existing tissues.

Bionanotechnology: The use of biological principles and building blocks for nanotechnological applications.

Biosensor: A device that combines a biological detection component together with a transducer component for the electrical or optical detection of analytes.

Biotechnology: The technological application of biology in the field of medicine and agriculture.

Biotin: A small organic compound of egg also known as vitamin H or B7 that forms a very strong complex with avidin.

Bottom up: A process that is based on the formation of complex structures by the coordinated assembly of simple building blocks (see also *top down*).

Cancer: A class of diseases or disorders that is characterized by the uncontrolled division of cells and formation of metastases.

Composite material: An engineered material that is composed of two or more constituent materials with significantly different physical or chemical properties and which remain separate and distinct within the finished structure.

Contrast agent: An agent that increases the sensitivity of a given imaging technique that improving the noise to signal ratio.

Copolymer: A polymer formed by two or more different types of monomer that are linked in the same chain.

Covalent bond: A chemical bond that is characterized by the sharing of one or more electrons between two atoms.

DNA: Deoxyribonucleic acid, a nucleic acid that contains the information for the synthesis of RNA molecules that is most cases are being translated into proteins.

Double helix: A supramolecular structure composed of two congruent helices with the same axis, differing by a translation along the axis. DNA has a double helix structure.

Drug delivery: A method for the delivery of a pharmaceutical compound to humans or animals.

Electron microscopy (EM): A type of microscope that uses electrons instead of photons to create an image of the target. The short wavelength of the electron beam allows nanometric resolution as compared to hundreds of nanometers resolution using optical microscopy.

Electroless deposition: A method to reduce metal ions into element metal by the use of chemical reducing agents.

Focused ion beam (FIB): An instrument that uses a focused beam of ions (usually gallium ions) for imaging and lithography at nano-scale resolution.

Graphite: A two dimensional flat sheets of carbon (from the Greek: to draw).

Homology: (in its biochemical context): The share of high degree of sequence identity or similarity in the DNA or protein levels.

Hydrogel: A network of polymer chains which contains predominantly water (typically more than 99%). Many hydrogels show nano-scale order.

Hydrophilic: A polar chemical compound that prefer polar solvent such as water (from the Greek *hydros*: water and *philia*: love). See: *Amphiphilic*.

Hydrophobic: A non-polar chemical compound which is repelled from a mass of water and prefer non-polar solvents (from the Greek *hydros*: water and *phobos*: fear). Hydrophobic compounds are sometimes known as "lipophilic".

Immunoglobulin: See *antibody*.

Inorganic: A chemical compound that do not contain carbon.

Ion Channel: A membrane embedded protein-made pore that allow the transport of specific ions (e.g., sodium channel or potassium channel).

Kevlar®: A brand name for a particular light but very strong aramid (aromatic polyamide) fibre (see also *protein*, *peptide*, *polyamide*). Trend name of DuPont.

Kinesin: A class of motor protein that moves along the microtubules to transport cellular cargo (also see *microtubule*).

Lab-on-a-chip: A device that is capable of handling extremely small fluid volumes down to less than pico liters of chemical analysis of analytes.

Lipophilic: see *hydrophobic*.

Liposome: A spherical phospholipids membrane vesicle with a diameter of tens to hundreds of nm.

Lithography: In general, a method for printing on a smooth surface. In the context of micro- and nano-electronics, a method for the patterning of substrates using a mask and source of radiation (that could be photonic, electronic, or ionic).

Microtubule: A protein nano-fiber with a diameter of about 25 nm and length varying from several micrometers to millimeters in axons of nerve cells.

Molar: A concentration unit. Each molar is composed of one mole ($\sim 6 \times 10^{23}$, Avogadro's number) of atoms of molecules in one liter of solvent.

Mole: See *molar*.

Molecular biology: The study of biology at the level of molecules by the use of tools from chemistry, biochemistry, and biophysics.

Molecular recognition: The strong and specific non-covalent interactions between molecules that allow the formation of stable supramolecular complex (also see *Supramolecular Chemistry*).

Monomer: A small molecule that could be chemically bonded to other monomers to form a polymer.

Myosin: A motor proteins found in eukaryotic tissues that moves along the actin filaments (also see *actin filaments*).

Nanometer (*abbr* nm): 1 billionth or 10^{-9} meter.

Nanoagriculture: The use of nanotechnology of agriculture purposes, including animal and plant applications.

Nanobiotechnology: The use of nanotechnology for biological and medical applications.

Nanocosmetics: The use of nanotechnology for cosmetic application such as delivery of cosmetically active compounds in liposomes or other nano-scale carriers.

Nanotubes: A tubular structure at the nano-scale could be composed of carbon, inorganic, or organic materials (also see *inorganic, organic*).

Non-covalent interaction: Chemical interaction that do not include sharing of electrons between atoms. Non-covalent interactions include hydrogen bonds, hydrophobic interaction, and van der Waals interactions (also see *covalent interactions*).

Nucleotide: The main constitutes of DNA and RNA, a chemical compound that consists of a heterocyclic base, a sugar, and one or more phosphate groups.

Nucleic acid: Polymers composed of nucleotides, either *DNA* or *RNA* (also see *DNA, nucleotide, RNA*).

Photon: The elementary particle responsible for electromagnetic phenomena.

Picomolar: A concentration of 10^{-12} Molar.

Peptide: Short polymers composed of two or more amino-acids (also see *protein*).

Peptide nucleic acid (PNA): A polymer similar to DNA or RNA but differing in the presence of an amide backbone.

Phospholipid: A class of lipids that contains a negatively-charged phosphate group. The main constitute of biological membranes.

Polyamide: A polymer that contains monomers joined by peptide (amide) bonds. Polyamide includes the natural protein and peptide but also the artificial Nylon, Kevlar®, and peptide nucleic acids (also see *protein, peptide, Kevlar*).

Polymer: A large molecule consisting of repeating structural units, or monomers, connected by covalent chemical bonds (see also *covalent bond*).

Polysaccharide: A polymers composed of monosaccharides joined together by glycosidic links.

Protein: A long polymer composed of 40-50 amino-acids or more. Large proteins can contain thousands of amino-acids (see also *peptide*).

Quantum dot: A semiconductor nanostructure that have unique optical and fluorescent properties.

Ribosome: An organelle in cells composed of RNA and proteins that allows the synthesis of proteins.

RNA: Ribonucleic acid, a nucleic acid polymer consisting of nucleotide monomers. mRNA is transcribed from the DNA to serve as the code for the synthesis of proteins. RNA can also serve as a structural and regulatory component.

S-layer: A cell membrane structure composed of protein and glycoprotein that is found in bacteria and archaea.

Self-Assembly or **Self-Association:** A process by which simple building blocks interact with each other to form a complex of higher complexity.

Small unilamellar vesicle (SUV): See *liposome*.

Supramolecular chemistry: A field of chemistry which is concerned with the formation of non-covalent chemical complexes, supramolecules.

Surfactant: A wetting agents that lower the surface tension of a liquid (from *surface active agent*). Surfactants are amphipathic in nature soluble in both polar and non-polar solvent.

Tissue engineering: The ability to engineer biological systems at the level of a tissue rather than cellular or sub-cellular.

Top-down: A process that is based on the miniaturization of complex system by sizing-down of its components (also see *bottom up*).

Virus: A parasite composed of DNA or RNA molecule that is wrapped in a protein capsid or lipid envelope (also see *bacteriophage*).

Water Technology or **WaTech:** Technology that is associated with the treatment of water such as desalination and sewage treatment.

Bibliography

Aggeli, A., Bell, M., Boden, N., Keen, J. N., Knowles, P. F., McLeish, T. C., Pitkeathly, M., and Radford, S. E. (1997) Responsive gels formed by the spontaneous self-assembly of peptides into polymeric beta-sheet tapes. *Nature* **386**, 259-262.

Aizenberg, J., Weaver, J. C., Thanawala, M. S., Sundar, V. C., Morse, D. E., and Fratzl, P. (2005) Skeleton of Euplectella *sp.*: Structural hierarchy from the nanoscale to the macroscale. *Science* **309**, 275-278.

Altman, M., Lee, P., Rich, A., and Zhang, S. (2000) Conformational behavior of ionic self-complementary peptides. *Protein Sci.* **9**, 1095-1105.

Banerjee, I. A., Yu, L., and Matsui, H. (2003) Cu nanocrystal growth on peptide nanotubes by biomineralization: size control of Cu nanocrystals by tuning peptide conformation. *Proc. Natl. Acad. Sci. USA* **100**, 14678-14682.

Bong, D. T., Clark, D. T., Granja, J. R., and Ghadiri, M. R. (2001) Self-assembling organic nanotubes. *Angew. Chem. Int. Ed.* **40**, 988-1011.

Braun, E., Eichen, Y., Sivan, U., and Ben-Yoseph, G. (1998) DNA-templated assembly and electrode attachment of a conducting silver wire. *Nature* **391**, 775-778.

Binning, G., Rohrer, H., Gerber, Ch. and Weibel, E. (1982) Surface Studies by Scanning Tunneling Microscopy. *Phys. Rev. Lett.* **49**, 57-61

Bini, E., Knight, D. P., and Kaplan, D. L. (2004) Mapping domain structures in silks from insects and spiders related to protein assembly. *J. Mol. Biol.* **335**, 27-40.

Brott, L. L., Naik, R. R., Pikas, D. J., Kirkpatrick, S. M., Tomlin, D. W., Whitlock, P. W., Clarson, S. J., and Stone, M. O. (2001) Ultrafast holographic nanopatterning of biocatalytically formed silica. *Nature* **413**, 291-293.

Chapman, M. R., Robinson, L. S., Pinkner, J. S., Roth, R., Heuser, J., Hammar, M., Normark, S., and Hulgren, S. J. (2002) Role of *Escherichia coli* curli operons in directing amyloid fiber formation. *Science* **295**, 851-855.

Cheng, X., Wang, Y., Hanein, Y., Bohringer, K. F., and Ratner, B. D. (2004) Novel cell patterning using microheater-controlled thermoresponsive plasma films. *J. Biomed. Mater. Res.* A. **70**, 159-168.

Cole, L. A. (1997) Immunoassay of human chorionic gonadotropin, its free subunits, and metabolites. *Clin. Chem.* **43**, 2233-2243.

Collins, P. G., Arnold, M. S., and Avouris, P. (2001) Engineering carbon nanotubes and nanotube circuits using electrical breakdown. *Science* **292**, 706-709.

DeLong, E. F., and Pace, N. R. (2001) Environmental diversity of bacteria and archaea. *Syst. Biol.* **50**, 470-478.

de Pablo, P. J., Schaap, I. A., MacKintosh, F. C., and Schmidt, C. F. (2003) Deformation and collapse of microtubules on the nanometer scale. *Phys. Rev. Lett.* **91**, 098101.

Djalali, R., Chen, Y.F., and Matsui, H. (2003) Au nanocrystal growth on nanotubes controlled by conformations and charges of sequenced peptide templates. *J. Am. Chem. Soc.* **125**, 5873-5879.

Douglas, T. (2003) Materials science. A bright bio-inspired future. *Science* **299**, 1192-1193.

Drexler, E. K. (1981) Molecular engineering: An approach to the development of general capabilities for molecular manipulation. *Proc. Natl. Acad. Sci. USA* **78**, 5275-5278.

Drexler, E. K. (1986) Engines of Creation: The Coming Era of Nanotechnology. Anchor Books. (available also freely online: http://www.foresight.org/EOC/)

Drexler, E. K. (1992) Nanosystems: molecular machinery, manufacturing, and computation. Wiley Interscience.

Eigler, D. M., and Schweizer, E. K. (1990) Positioning single atoms with a scanning tunneling microscope. *Nature* **344**, 524-526.

Elliot, M. A., Karoonuthaisiri, N., Huang, J., Bibb, M. J., Cohen, S. N., Kao, C. M., and Buttner, M. J. (2003) The chaplins: a family of hydrophobic cell-surface proteins involved in aerial mycelium formation in Streptomyces coelicolor. *Genes Dev.* **17**, 1727-1740.

Empedocles, S. A. and Bawendi, M. G. (1997) Quantum-confined stark effect in single CdSe nanocrystallite quantum dots. *Science* **278**, 2114-2117.

Gazit, E. (2002) A possible role for pi-stacking in the self-assembly of amyloid fibrils. *FASEB J.* **16**, 77-83.

Gazit, E. (2002) Mechanistic studies of the process of amyloid fibrils formation by the use of peptide fragments and analogues: Implications for the design of fibrillization inhibitors. *Curr. Med. Chem.* **9**, 1725-1735.

Ghadiri, M. R., Granja, J. R., Milligan, R. A., McRee, D.E., and Khazanovich, N. (1993) Self-assembling organic nanotubes based on a cyclic peptide architecture. *Nature* **366**, 324-347.

Ghadiri, M. R., Granja, J. R., and Buehler, L. K. (1994) Artificial transmembrane ion channels from self-assembling peptide nanotubes. *Nature* **369**, 301-304.

Gilead, S., and Gazit, E. (2005) Self-organization of short peptide fragments: from amyloid fibrils to nanoscale supramolecular assemblies. *Supramol. Chem.* **17**, 87-92.

Goel, A., Howard, J. B., and Sande, J. B. V. (2004) Size analysis of single fullerene molecules by electron microscopy. *Carbon* **42**, 1907-1915.

Gyorvary, E., Schroedter, A., Talapin, D. V., Weller, H., Pum, D., and Sleytr, U. B. (2004) Formation of nanoparticle arrays on S-layer protein lattices. *J. Nanosci. Nanotechnol.* **4**, 115-120.

Favier, F., Walter, E.C., Zach, M.P., Benter, T., and Penner, R.M. (2001) Hydrogen sensors and switches from electrodeposited palladium mesowire arrays. *Science* **293**, 2227-2231.

Feldkamp, U., and Niemeyer, C. M. (2006) Rational design of DNA nanoarchitectures. *Angew. Chem. Int. Ed.* **45**, 1856-1876.

Fraenkel-Conrat, H., Singer, B., and Williams, R. C. (1957) Infectivity of viral nucleic acid. *Biochim. Biophys. Acta.* **25**, 87-96.

Foley, J., Schmid, H., Stutz, R., and Delamarche, E. (2005) Microcontact printing of proteins inside microstructures. *Langmuir* **21**, 11296-11303.

Fromherz, P. (2002) Electrical interfacing of nerve cells and semiconductor chips. *Chemphyschem* **3**, 276-284.

Hamada, D., Yanagihara, I., and Tsumoto, K. (2004) Engineering amyloidogenicity towards the development of nanofibrillar materials. *Trends Biotechnol.* **22**, 93-97.

Hart, S. L. (2005) Lipid carriers for gene therapy. *Curr. Drug Deliv.* **2**, 423-428.

Hartgerink, J. D., Beniash, E., and Stupp, S. I. (2001) Self-assembly and mineralization of peptide-amphiphile nanofibers. *Science* **294**, 1684-1688.

Holmes, T. C., de Lacalle, S., Su, X., Liu, G., Rich, A., and Zhang, S. (2000) Extensive neurite outgrowth and active synapse formation on self-assembling peptide scaffolds. *Proc. Natl. Acad. Sci. USA* **97**, 6728-6733.

Horne, W.S., Ashkenasy, N., and Ghadiri, M. R. (2005) Modulating charge transfer through cyclic D,L-alpha-peptide self-assembly. *Chemistry* **11**, 1137-1144.

Hunter, C. A. (1993) Arene-Arene Interactions: Electrostatic or charge transfer. *Angew. Chem. Int. Ed.* **32**, 1584-1586.

Iijima, S. (1991) Helical microtubules of graphitic carbon *Nature* **354**, 56-58.

Jaffari, S. A., and Turner, A. P. (1995) Recent advances in amperometric glucose biosensors for in vivo monitoring. *Physiol Meas.* **16**, 1-15

Jaremko, J., and Rorstad, O. (1998) Advances toward the implantable artificial pancreas for treatment of diabetes. *Diabetes Care* **21**, 444-450

Jarrett, J. T., and Lansbury, P. T. Jr. (1993) Seeding "one-dimensional crystallization" of amyloid: a pathogenic mechanism in Alzheimer's disease and scrapie? *Cell* **73**, 1055-1058.

Jin, H.-J., and Kaplan, D.E. (2003) Mechanism of silk processing in insects and spiders. *Nature* **424**, 1057-1061.

Kamiya, S., Minamikawa, H., Jung, J. H., Yang, B., Masuda, M., and Shimizu, T. (2005) Molecular structure of glucopyranosylamide lipid and nanotube morphology. *Langmuir* **21**, 743-750.

Kane, R. S., Takayama, S., Ostuni, E., Ingber, D. E., and Whitesides, G. M. (1999) Patterning proteins and cells using soft lithography. *Biomaterials* **20**, 2363-2376.

Katz, E., and Willner, I. (2004) Biomolecule-functionalized carbon nanotubes: applications in nanobioelectronics. *Chemphyschem* **5**, 1084-1104.

Kaul, R. A., Syed, N. I., and Fromherz, P. (2004) Neuron-semiconductor chip with chemical synapse between identified neurons. *Phys. Rev. Lett.* **92**, 038102.

Keren, K., Krueger, M., Gilad, R., Ben-Yoseph, G., Sivan, U., and Braun E. (2002) Sequence-specific molecular lithography on single DNA molecules. *Science* **297**, 72-75.

King, C. Y., Tittmann, P., Gross, H., Gebert, R., Aebi, M., and Wuthrich, K. (1997). Prion-inducing domain 2-114 of yeast Sup35 protein transforms *in vitro* into amyloid-like filaments. *Proc. Natl. Acad. Sci. USA* **94**, 6618-6622.

Kisiday, J., Jin, M., Kurz, B., Hung, H., Semino, C., Zhang, S., and Grodzinsky, A. J. (2002) Self-assembling peptide hydrogel fosters chondrocyte extracellular matrix production and cell division: implications for cartilage tissue repair. *Proc. Natl. Acad. Sci. USA* **99**, 9996-10001.

Kiley, P., Zhao, X., Vaughn, M., Baldo, M. A., Bruce, B. D., and Zhang, S. (2005) Self-assembling peptide detergents stabilize isolated photosystem I on a dry surface for an extended time. *PLoS Biol.* **3**, e230.

Knez, M., Sumser, M. P., Bittner, A. M., Wege, C., Jeske, H., Martin, T. P., and Kern, K. (2004). Spatially selective nucleation of metal clusters on the tobacco mosaic virus. *Adv. Funct. Mat.* **14**, 114-119.

Knez, M., Bittner, A. M., Boes, C., Wege, C., Jeske, H., Maiß, K., and Kern, K. (2003). Biotemplate synthesis of 3-nm nickel and cobalt nanowires. *Nano Lett.* **3**, 1079-1082.

Kol, N., Abramovich, L., Barlam, D., Shneck, R. Z., Gazit E., and Rousso, I. (2005) Self-assembled peptide nanotubes are uniquely rigid bioinspired supramolecular structures. *Nano Lett.* **5**, 1343-1346.

Kroto, H. W., Heath, J. R., O'Brien, S. C., Curl, R. F., and Smalley, R. E. (1985) C60: Buckminsterfullerene. *Nature* **318**, 162-163.

Kubik, S. (2002) High-performance fibers from spider silk. *Angew. Chem. Int. Ed.* **41**, 2721-2723.

Langer, R. (2000) Biomaterials in drug delivery and tissue engineering: One laboratory's experience. *Acc. Chem. Res.* **33**, 94-101.

Lee, S. W., Mao, C., Flynn, C. E., and Belcher, A. M. (2002a) Ordering of quantum dots using genetically engineered viruses. *Science* **296**, 892-895.

Lee, K.-B., Park, S.-J., Mirkin, C. A., Smith, J. C., and Mrksich, M. (2002b) Protein nanoarrays generated by dip-pen nanolithography. *Science* **295**, 1702-1705.

Lee, K-B., Lim, J-H., and Mirkin, C.A. (2003) Protein Nanostructures Formed Via Direct-Write Dip-Pen Nanolithography. *J. Am. Chem. Soc.* **125**, 5588-5589.

Lehn, J. M. (2002) Toward self-organization and complex matter. *Science* **295**, 2400-2403.

Li Jeon, N., Baskaran, H., Dertinger, S. K., Whitesides, G. M., Van de Water, L., and Toner, M. (2002) Neutrophil chemotaxis in linear and complex gradients of interleukin-8 formed in a microfabricated device. *Nat. Biotechnol.* **20**, 826-830.

Luckey, M., Hernandez, J., Arlaud, G., Forsyth, V. T., Ruigrok, R. W., and Mitraki, A. (2000) A peptide from the adenovirus fiber shaft forms amyloid-type fibrils. *FEBS Lett.* **468**, 23-27.

Liu, D., Park, S. H., Reif, J. H., and LaBean, T. H. (2004) DNA nanotubes self-assembled from triple-crossover tiles as templates for conductive nanowires. *Proc. Natl. Acad. Sci. USA* **101**, 717-722.

Liu, Z., Zhang, D., Han, S., Li, C., Lei, B., Lu, W., Fang, J., and Zhou, C. (2005) Single Crystalline Magnetite Nanotubes. *J. Am. Chem. Soc.* **127**, 6-7.

Madhavaiah, C., and Verma, S. (2004) Self-aggregation of reverse bis peptide conjugate derived from the unstructured region of the prion protein. *Chem. Commun.* **21**, 638-639.

Madhavaiah, C., and Verma, S. (2005) Copper-metalated peptide palindrome derived from prion octarepeat: synthesis, aggregation, and oxidative transformations. *Bioorg. Med. Chem.* **13**, 3241-3248.

Magrisso, M., Etzion, O., Pilch, G., Novodvoretz, A., Perez-Avraham, G., Schlaeffer, F., and Marks R. (2006) Fiber-optic biosensor to assess circulating phagocyte activity by chemiluminescence. *Biosens. Bioelectron.* **21**, 1210-1218.

Mao, C., Solis, D. J., Reiss, B. D., Kottmann, S. T., Sweeney, R. Y., Hayhurst, A., Georgiou, G., Iverson, B., and Belcher, A. M. (2004) Virus-based toolkit for the directed synthesis of magnetic and semiconducting nanowires. *Science* **303**, 213-217.

Mathe, J., Aksimentiev, A., Nelson, D. R., Schulten, K., and Meller, A. (2005) Orientation discrimination of single-stranded DNA inside the alpha-hemolysin membrane channel. *Proc. Natl. Acad. Sci. USA* **102**, 12377-12382.

Nakashima, N., Asakuma, S., and Kunitake, T. (1985) Optical microscopic study of helical superstructures of chiral bilayer membranes. *J. Am. Chem. Soc.* **107**, 509-510.

Nam, J-M., Thaxton, C. S., and Mirkin, C. A. (2003) Nanoparticle-based bio-bar codes for the ultrasensitive detection of proteins. *Science* **301**, 1884-1886.

Nam, K. T., Kim, D. W., Yoo, P. J., Chiang, C. Y., Meethong, N., Hammond, P. T., Chiang, Y. M., and Belcher, A. M. (2006) Virus-enabled synthesis and assembly of nanowires for lithium ion battery electrodes. *Science* **312**, 885-888.

Mann, S., and Weiner, S. (1999) Biomineralization: Structural questions at all length scales. *J. Struct. Biol.* **126**, 179-181.

Margulis, L., Salitra, G., Tenne, R., and Talianker, M. (1993) Nested fullerene-like structures. *Nature* **365**, 113-114.

Martin, R., Waldmann, L., and Kaplan, D. L. (2003) Supramolecular assembly of collagen triblock peptides. *Biopolymers* **70**, 435-444.

Marx, A., Muller, J., and Mandelkow, E. (2005) The structure of microtubule motor proteins. *Adv. Protein Chem.* **71**, 299-344.

Meller, A., Nivon, L., Brandin, E., Golovchenko, J., and Branton, D. Rapid nanopore discrimination between single polynucleotide molecules. *Proc. Natl. Acad. Sci. USA* **97**, 1079-1084.

Merrifield R. B. (1965) Automated synthesis of peptides. *Science* **150**, 178-185.

Mitraki, A., and van Raaij, M. J. (2005) Folding of beta-structured fibrous proteins and self-assembling peptides. *Methods Mol. Biol.* **300**, 125-140.

Morais-Cabral, J. H., Lee, A., Cohen, S. L., Chait, B. T., Li, M., and Mackinnon, R. (1998) Crystal structure and functional analysis of the HERG potassium channel N terminus: a eukaryotic PAS domain. *Cell* **95**, 649-655.

Naik, R. R., Whitlock, P. W., Rodriguez, F., Brott, L. L., Glawe, D. D., Clarson, S. J., and Stone, M. O. (2003) Controlled formation of biosilica structures *in vitro*. *Chem. Commun.* **2003**, 238-239.

Nielsen, P. E. (1995) DNA analogues with nonphosphodiester backbones. Annu Rev *Biophys. Biomol. Struct.* **24**, 167-183.

Nowak, A. P., Breedveld, V., Pakstis, L., Ozbas, B., Pine, D. J., Pochan, D., Deming, T. J. (2002) Rapidly recovering hydrogel scaffolds from self-assembling diblock copolypeptide amphiphiles. *Nature* **417**, 424-428.

Nowak, A. P., Breedveld, V., Pine, D. J., and Deming, T. J. (2003) Unusual salt stability in highly charged diblock co-polypeptide hydrogels. *J. Am. Chem. Soc.* **125**, 15666-15670.

Nuraje, N., Banerjee, I. A., MacCuspie, R. I., Yu, L., and Matsui, H. (2004) Biological bottom-up assembly of antibody nanotubes on patterned antigen arrays. *J. Am. Chem. Soc.* **126**, 8088-8089.

Papanikolopoulou, K., Schoehn, G., Forge, V., Forsyth, V. T., Riekel, C., Hernandez, J. F., Ruigrok, R. W., and Mitraki, A. (2005) Amyloid fibril formation from sequences of a natural beta-structured fibrous protein, the adenovirus fiber. *J. Biol. Chem.* **280**, 2481-2490.

Parker, M. W., and Feil, S. C. (2005) Pore-forming protein toxins: from structure to function. *Prog. Biophys. Mol. Biol.* **88**, 91-142.

Patolsky, F., Weizmann, Y., and Willner, I. (2004) Actin-based metallic nanowires as bio-nanotransporters. *Nat. Mater.* **3**, 692-695.

Patolsky, F., Zheng, G., Hayden, O., Lakadamyali, M., Zhuang, X., and Lieber C. M. (2004) Electrical detection of single viruses. *Proc. Natl. Acad. Sci U S A.* **101**, 14017-14022.

Perry, C. C., and Keeling-Tucker, T. (2000) Biosilicification: the role of the organic matrix in structure control. *J. Biol. Inorg. Chem.* **5**, 537-550.

Pfeffer, R. (1927) *Organische Molekülverbindungen.* Enke, Stuttgart.

Porat, Y., Kolusheva, S., Jelinek, R., and Gazit, E. (2003) The human islet amyloid polypeptide forms transient membrane-active protofilaments. *Biochemistry* **42**, 10971-10977.

Reches, M, and Gazit, E. (2003) Casting metal nanowires within discrete self-assembled peptide nanotubes. *Science* **300**, 625-627.

Reches, M., and Gazit, E. (2004) Formation of closed-cage nanostructures by self-assembly of aromatic dipeptides. *Nano Lett.* **4**, 581-585.

Reches, M., and Gazit, E. (2005) Self-assembly of peptide nanotubes and amyloid-like structures by charged-termini capped diphenylalanine peptide analogues. *Israel J. Chem.* **45**, 363-371.

Reches, M., and Gazit, E. (2006) Molecular self-assembly of peptide nanostructures: mechanism of association and potential uses. *Curr. Nanoscience* **2**, 105-111.

Schinazi, R. F., Sijbesma, R., Srdanov, G., Hill, C. L., and Wudl, F. (1993) Synthesis and virucidal activity of a water-soluble, configurationally stable, derivatized C60 fullerene. *Antimicrob. Agents Chemother.* **37**, 1707-1710.

Sahoo,Y., Pizem, H., Fried, T., Goldnitsky, D., Burstein, L., Sukenik, C.M., and Markovich, G. (2001) Alkyl Phosphonate/Phosphate Coating on Magnetite Nanoparticles: A Comparison with Fatty Acids. *Langmuir* **17**, 7907-7911.

Sarikaya, M., Tamerler, C., Jen, A.K., Schulten, K., and Baneyx, F. (2003) Molecular biomimetics: nanotechnology through biology. *Nat Mater.* **2**, 577-585.

Scheibel, T., Parthasarathy, R., Sawicki, G., Lin, X. M., Jaeger, H., and Lindquist, S. L. (2003) Conducting nanowires built by controlled self-assembly of amyloid fibers and selective metal deposition. *Proc. Natl. Acad. Sci. USA* **100**, 4527-4532.

Schneider, J. P., Pochan, D. J., Ozbas, B., Rajagopal, K., Pakstis, L., and Kretsinger, J. Responsive hydrogels from the intramolecular folding and self-assembly of a designed peptide. *J. am. Che. Soc.* **124**, 15030-15037.

Schuster, B., Gyorvary, E., Pum, D., and Sleytr, U. B. (2005) Nanotechnology with S-layer proteins. Nanotechnology with S-layer proteins. *Methods Mol. Biol.* **300**, 101-123.

Seeman, N. C. (1990) De novo design of sequences for nucleic acid structural engineering. *J. Biomol. Struct. Dyn.* **8**, 573-581.

Seeman, N. C. (2005) Structural DNA nanotechnology: an overview. *Methods Mol. Biol.* **303**, 143-166.

Sherman, W. B., and Seeman, N. C. (2006) Design of minimally strained nucleic Acid nanotubes. *Biophys. J.* **90**, 4546-4557.

Shim, J., Bersano-Begey, T. F., Zhu, X., Tkaczyk, A. H., Linderman, J. J., and Takayama, S. (2003) Micro- and nanotechnologies for studying cellular function. *Curr. Top. Med. Chem.* **3**, 687-703.

Silva, G. A., Czeisler, C., Niece, K. L., Beniash, E., Harrington, D. A., Kessler, J. A., and Stupp, S. I. (2004) Selective differentiation of neural progenitor cells by high-epitope density nanofibers. *Science* **303**, 1352-1355.

Simon, P., Lichte, H., Wahl, R., Mertig, M., and Pompe W. (2004) Electron holography of non-stained bacterial surface layer proteins. *Biochim. Biophys. Acta.* **1663**, 178-187.

Sun, S., and Bernstein, E. R. (1996) Aromatic van der Waals Clusters: Strcture and nonrigidity. *J. Phys. Chem.* **100**, 13348-13366.

Sundar, V. C., Yablon, A. D., Grazul, J. L., Ilan, M., and Aizenberg, J. (2003) Fibre-optical features of a glass sponge. *Nature* **424**, 899-900.

Taniguchi, N. (1974) On the basic concept of 'NanoTechnology' *Proc. Intl. Conf. Prod. Eng. Tokyo, Part II, Japan Society of Precision Engineering.*

Tenidis, K., Waldner, M., Bernhagen, J, Fischle, W., Bergmann, M., Weber, M., Merkle, M. L., Voelter, W., Brunner, H., and Kapurniotu, A. Identification of a penta- and hexapeptide of islet amyloid polypeptide (IAPP) with amyloidogenic and cytotoxic properties. *J. Mol. Biol.* **295**, 1055-1071.

Tenne, R. (2002) Inorganic nanotubes and fullerene-like materials. *Chemistry* **8**, 5296-5304.

Tenne, R., and Rao, C.N. (2004) Inorganic nanotubes. *Philos. Transact. A. Math. Phys. Eng. Sci.* **362**, 2099-2125.

True, H. L., Berlin, I., and Lindquist, S. L. (2004) Epigenetic regulation of translation reveals hidden genetic variation to produce complex traits. *Nature* **431**, 184-187.

Valery, C., Paternostre, M., Robert, B., Gulik-Krzywicki, T., Narayanan, T., Dedieu, J. C., Keller, G., Torres, M. L., Cherif-Cheikh, R., Calvo, P., and Artzner, F. (2003) Biomimetic organization: Octapeptide self-assembly into nanotubes of viral capsid-like dimension. *Proc. Natl. Acad. Sci. USA* **100**, 10258-10262.

Vauthey, S, Santoso, S., Gong, H., Watson, N., and Zhang, S. (2002) Molecular self-assembly of surfactant-like peptides to form nanotubes and nanovesicles. *Proc. Natl. Acad. Sci. USA* **99**, 5355-5360.

Wang, W. U., Chen, C., Lin, K. H., Fang, Y., and Lieber, C. M. (2005) Label-free detection of small-molecule-protein interactions by using nanowire nanosensors. *Proc. Natl. Acad. Sci. USA* **102**, 3208-3212.

Wang, Y. C., and Ho, C. C. (2004) Micropatterning of proteins and mammalian cells on biomaterials. *FASEB J.* **18**, 525-527.

Weiner, S., Sagi, I., and Addadi, L. (2005) Structural biology. Choosing the crystallization path less traveled. *Science* **309**, 1027-1028.

Wilchek, M, and Bayer, E. A. (1990) Introduction to avidin-biotin technology. *Methods Enzymol.* **184**, 5-13.

Whitesides, G.M., Mathias, J.P., and Seto, C.T. (1991) Molecular self-assembly and nanochemistry: A chemical strategy for the synthesis of nanostructures. *Science* **254**, 1312-1319.

Whitesides, G. M., Ostuni, E., Takayama, S., Jiang, X., and Ingber, D. E. (2001) Soft lithography in biology and biochemistry. *Annu. Rev. Biomed. Eng.* **3**, 335-373.

Wu, H., Wheeler, A., Zare, R. N. (2004) Chemical cytometry on a picoliter-scale integrated microfluidic chip. *Proc. Natl. Acad. Sci. USA* **101**, 12809-12813.

Xiao, Y., Patolsky, F., Katz, E., Hainfeld, J. F., and Willner, I. (2003) "Plugging into Enzymes": Nanowiring of redox enzymes by a gold nanoparticle. *Science* **299**, 1877-1881.

Yager, P., and Schoen, P. E. (1984) Formation of tubules by a polymerizable surfactant. *Mol. Cryst. Liq. Cryst.* **106**, 371-381.

Yamada, K., Ihara, H., Ide, T., Fukumoto, T., and Hirayama, C. (1984) Formation of helical super structure from single-walled bilayers by amphiphiles with oligo-L-glutamic acid-head group. *Chem. Lett.* **1984**, 1713-1716.

Yan, H., LaBean, T. H., Feng, L., and Reif, J. H. (2003) Directed nucleation assembly of DNA tile complexes for barcode-patterned lattices. *Proc. Natl. Acad. Sci. USA* **100**, 8103-8108.

Yemini, M., Reches, M., Rishpon, J., and Gazit, E. (2005a) Novel electrochemical biosensing platform using self-assembled peptide nanotubes. *Nano Lett.* **5**, 183-186.

Yemini, M., Reches, M., Gazit, E., and Rishpon, J. (2005b) Peptide nanotubes modified electrodes for enzyme-biosensors applications. *Anal. Chem.* **77**, 5155-5159.

Zhang, S. (2003) Fabrication of novel biomaterials through molecular self-assembly. *Nat Biotechnol.* **21**, 1171-1178.

Zheng, G., Patolsky, F., Cui, Y., Wang, W. U., and Lieber, C. M. (2005) Multiplexed electrical detection of cancer markers with nanowire sensor arrays. *Nat. Biotechnol.* **23**, 1294-1301.

Index